New Land, New Life: A success story of new land resettlement in Bangladesh

New Land, New Life: A success story of new land resettlement in Bangladesh

Editors:

Natasha Haider, *Project Manager/Consultant, Char Development and Settlement Project IV, Euroconsult Mott MacDonald*

Andrew Jenkins, *Technical Assistance Team Leader, Char Development and Settlement Project IV*

Under the coordination of Benoît Thierry

CABI

CABI is a trading name of CAB International

CABI
Nosworthy Way
Wallingford
Oxfordshire OX10 8DE
UK

Tel: +44 (0)1491 832111
Fax: +44 (0)1491 833508
E-mail: info@cabi.org
Website: www.cabi.org

CABI
WeWork
One Lincoln St
24th Floor
Boston, MA 02111
USA

T: +1 (617)682-9015
E-mail: cabi-nao@cabi.org

A catalogue record for this book is available from the British Library, London, UK.

Library of Congress Cataloging-in-Publication Data

Names: Jenkins, Andrew, editor. | Haider, Natasha, editor.
Title: New land, new life : a success story of new land resettlement in Bangladesh / Andrew Jenkins, Natasha Haider.
Description: Boston, MA : CAB International, [2020] | Includes bibliographical references. | Summary: "The Ganges-Brahmaputra-Meghna deltas has newly emerged 'char' islands, which are very vulnerable, socially, institutionally and environmentally. This book explains how the Governments of Bangladesh and the Netherlands cooperated on a land-based rural development project to give settlers security and purpose"-- Provided by publisher.
Identifiers: LCCN 2020018112 | ISBN 9781789246049 (paperback) | ISBN 9781789246056 | ISBN 9781789246063
Subjects: LCSH: Land settlement--Bangladesh.
Classification: LCC HD1131 .N49 2020 | DDC 333.3/15492--dc23
LC record available at https://lccn.loc.gov/2020018112

References to Internet websites (URLs) were accurate at the time of writing.

ISBN-13: 9781789246049 (pbk)

Commissioning editor: David Hemming
Production editor: Shankari Wilford

Printed and bound in the UK from copy supplied by the authors by Severn, Gloucester

Contents

Figures and Tables

Figures

Tables

Contributors

Bazlul Karim, Deputy Team Leader, CDSP IV

Mihir Chakraborty, Deputy Team Leader, CDSP IV

Kiran Sankar Sarker, Monitoring and Evaluation Specialist, CDSP IV

Md. Rezaul Karim, Land Advisor, CDSP IV

Md. Robiul Islam, Social Forestry Adviser, CDSP IV

Nujulee Begum, Gender Expert, CDSP III

Edward Mallorie, Economist, CDSP IV

Koen de Wilde, Team Leader, CDSP II & III

Glossary

AC (Land)	Assistant Commissioner (Land)
AOS	Annual Outcome Surveys
BRAC	Building Resources Across Community
BWDB	Water Development Board
CDSP	Char Development & Settlement Project
CO	Credit Officer
DAE	Department of Agricultural Extension
DFID	The Department for International Development
DLS	Department of Livestock Services
DPHE	Department of Public Health Engineering
DPP	Development Project Proforma
DTW	Deep Tube Well
DUS	Dwip Unnayan Sangsthan
DAE	Department Of Agriculture Extension
EKN	Embassy of the Kingdom of the Netherlands
FD	Forest Department
FF	Farmers Forum
FLIs	Field Level Institutions
GAP	Gender Action Plan
GoB	Government of Bangladesh
GoN	Government of the Netherlands
HH	Household
HYV	High-yielding variety
ICC	Inter Agency Coordination Committee
ICZM	Integrated Coastal Zone Management
IFAD	International Fund for Agricultural Development
IGAs	Income Generating Activities
LCS	Labour Contracting Societies

LGED	Local Government Engineering Department
LGIs	Local Government Institutions
LHR	Legal Human Rights
LRMS	Land Records Management System
LRP	Land Reclamation Project
MES	Meghna Estuary Study
MFI	Micro Finance Institution
MoL	Ministry of Land
MoWR	Ministry of Water Resources
MWR	Ministry of Water Resources
O&M	Operation & Maintenance
PDO	Project Development Office
PKSF	Polli Karma Sahayak Foundation
PMC	Project Management Committee
PNGO	Partner Non-Government Organization
PTPS	plot-to-plot-survey
PWD	Public Works Department
PWM	Participatory Water Management
RFLDC	Regional Fisheries and Livestock Development Component
SDI	Society for Development Initiatives
SFG	Social Forestry Group
SSUS	Sagorika Samaj Unnayan Sangsthan
SLS	Social & Livelihood Support
TUG	Tube Well User Group
TA	Technical Assistance
TBA	Traditional Birth Attendant
TUG	Tubewell User Group
UMDC	Union Disaster Management Committees
UP	Union Parishad
WARPO	Water Resources Planning Organization

WATSAN	Water and Sanitation
WMA	Water Management Association
WMF	Water Management Federation
WMG	Water Management Groups
WMO	Water Management Organization

Conversions

Area: 100 decimal = 1 acre = 0.4 hectares

Currency: Bangladesh Taka (BDT): Tk.83 = USD 1, Tk.101 = EUR 1

Acknowledgements

The chapters in this book are based on experience gathered from implementing the Char Development and Settlement Project in Bangladesh, being undertaken since 1994. The Governments of Bangladesh and the Netherlands cooperated to work on char development and settlement, starting with the Netherlands-supported Land Reclamation Project (LRP) in 1977.

CDSP-IV, implemented from 2011 to 2018, was implemented by the Government of Bangladesh, supported by the United Nations International Fund for Agricultural Development (IFAD), and the Government of the Netherlands.

Important contributions were made by:
Implementing Government Agencies: Bangladesh Water Development Board (BWDB); Ministry of Land (MoL); Local Government Engineering Department (LGED); Department of Public Health Engineering (DPHE); Department of Agricultural Extension (DAE); Forest Department (FD)
PNGOs: BRAC; Dwip Unnayan Sangstha (DUS); Society for Development Initiatives (SDI).
Technical Assistance: Euroconsult Mott MacDonald; BETS; Socioconsult.

We wish to recognize the project technical assistance team members: Andrew Jenkins, Bazlul Karim, Mihir Chakraborty, Kiran Sankar Sarker, Md. Rezaul Karim, Md. Robiul Islam, Nujulee Begum, and Edward Mallorie for their input and advice on project activities and outputs.

We are particularly grateful to: Mr. Md. Shamsuddoha, Project Coordinating Director, PCD-CDSP-IV; Mr. Peter de Vries, First Secretary Embassy of the Kingdom of the Netherlands, Bangladesh; Mr Benoît Thierry, former Senior Country Programme Manager at IFAD; Mr Omer Zafar, Country Director, Bangladesh and Maldives, IFAD; Ms. Sherina Tabassum, Country Programme Officer, Bangladesh, IFAD; and Dr. Hero Heering, Projects Director, Euroconsult Mott MacDonald for their support in developing this book.

Chapter 1

The Coastal Chars of Bangladesh

Excerpt from Moving Coastlines: Emergence and Use of Land in the Ganges-Brahmaputra-Meghna Estuary (Wilde, 2011)

The Ganges-Brahmaputra-Meghna can be considered to be one of the most dynamic deltas in the world. This is because the river system carries sediments originating from distant northern India and the Tibetan Plateau into the Bay of Bengal, creating a permanent process of accretion and erosion. In deltas and estuaries, the deposition of sediment is carried out by river or supplied from the sea by tidal action. The net deposition results in the growth of a delta in the estuary as newly emerged islands known as 'Chars' in Bangla. This process of land accretion is a continuous and very slow natural process maintaining land elevation and soil fertility. For the purpose of this book we are focusing on the Meghna Estuary which is the only active delta-forming estuary in coastal Bangladesh. The estuary is part of the coastal zone of Bangladesh and is being shaped by a very complex set of interactions between physical processes. Some of the key factors having a long-term effect in the estuary are: shifting of the river mouths; changing of the base level; natural hazards; and climate change. Total annual sediment discharge into the lower Meghna is on average about 1,100 million tons per year, of which about one fifth is retained in the Meghna estuary. This forms the material for land accretion in the central part of the coastal zone (Wilde, 2011).

In the Meghna Estuary the rate of net land accretion is approximately 25 km^2 per year which is much higher compared with that of the past. This further encourages efforts to accelerate the natural process by engineering interventions to gain more land which are beneficial for a country like Bangladesh, considering the population pressure. Accretion of land is also useful to cope with natural disasters (i.e. cyclones and storm surges) and climate change impacts such as sea level rise. In Bangladesh the process of land reclamation to accelerate natural accretion was started in 1956-57 by closure of channels and construction of cross-dams (Wilde, 2011). The Netherlands-supported Land Reclamation Project launched in 1977 was one of the first systematic efforts to study the potential for land reclamation. Since then, the Bangladesh Water Development Board has carried out a number of projects using cross dams to accelerate the build-up of land. As a result, the country has so far reclaimed over 1,000 km^2 of land from the sea, South of Noakhali District (Islam, 2015).

Char areas are physically different from other parts of Bangladesh. These areas are low lying and consequently vulnerable to flooding and cyclones from the Bay of Bengal. Soils of char areas are high in salinity and low in organic materials. The youngest chars are mud flats supporting little vegetation, dissected by tidal creeks, subject to frequent flooding during high tides. The oldest chars are already consolidated lands, supporting annual cropping and more or less permanent homesteads, despite the fact that the lands are unprotected and vulnerable to extensive crop damage from cyclones (Wilde, 2000).

People migrate to recently emerged chars for a variety of reasons, primarily, because people lost their original land and homesteads as a result of erosion (this is probably the case for 80-90% of the households). When a new char becomes fit for cultivation, the river-eroded families from adjacent areas start migrating into the newly formed land for shelter and livelihood. Small numbers of families living in close association form a type of community called *Samaj* in Bangla. *Samaj* gives people a sense of security. At this stage, service delivery mechanisms from government agencies are hardly present in char areas and the private sector is usually limited to small shops. Non-Governmental Organisations (NGOs) are generally present but in less density than in other parts of Bangladesh. As a result, in the absence of a formal institutional network, a power broker, in many cases with ancestral links to newly accreted char land, tends to extend support and patronage to settlers. This type of autonomous settlement leads to a situation in which the official process of land settlement cannot start with a clean slate. Settlers are already present in new chars with active control over land before the official process has even started. Powerful people, commonly known as *jotdar*, and the settlers controlled by them, occupy the land. The immigration is illegal, because the land is under the control of the Forest Department for a period of 20 years after the start of the first afforestation activities (Wilde, 2000).

The land is subject to regular flooding. There is very limited access to drinking water, especially in winter, and no system of communication. For food, the settlers are dependent on a low-yielding rice aman crop, some rabi crops and a few fish farmed in ponds or caught in open waters. Some income is derived from tending cattle. People have no official title for the land they occupy. They are vulnerable to a set of risks such as flooding, storms and salinity intrusion.

Chapter 2

The Birth of CDSP-IV

Kiran Sankar Sarker (Wilde, 2000)

To reduce the social, institutional and environmental vulnerability faced in char areas, development interventions were necessary to provide a sense of security at different levels and to unleash the development potential that the chars offer (Wilde, 2000). The Governments of Bangladesh and the Netherlands cooperated to work on char development and settlement, starting with the inception of the Netherlands-supported Land Reclamation Project (LRP) in 1977. During this project, which ended in 1991, the focus shifted from surveys and trials of land accretion to the development of new land. In order to continue both planning and land development activities, the LRP was then split into two separate projects: the Meghna Estuary Study (MES), for water-based surveys and trials, and the Char Development and Settlement Project (CDSP), a land-based rural development project. In an environment where vulnerability is the most prominent feature, additional investment by the settlers can only be expected if they are provided with a sense of security. Much of CDSP interventions are geared towards that purpose (Wilde, 2000).

Table 2.1. Chronology of main events

1975	Identification Mission commissioned by the Government of the Netherlands
May 1977	An Agreement arranging the technical cooperation between the Bangladesh Government and The Netherlands Government was signed
December 1977	LRP started formally
1981	Construction of a pilot polder at Char Baggar Dona started
1986	Settlement of landless in Char Baggar Dona started
November 1990	Appraisal Mission recommended the termination of LRP by mid-1991

September 1994	Char Development and Settlement Project (CDSP-I) started
September 1999	CDSP-II started (1999-2005)
February 2002	ICZM initiative[1] launched
July 2002	Mid Term Review Mission by the Netherlands recommended continuation of CDSP-II and extension of activities in Boyer Char
February 2004	Governments of Bangladesh and the Netherlands agreed the next phase
October 2005	CDSP-III starts for a period of 4 years and later extended for another 3 years (2005-2011)
March 2011	CDSP-IV starts (2011-2018)

The long-term development objective of the project is defined: ***'To bring about an improvement in the economic situation and in the living conditions of the coastal chars'*** and the project objectives are:

- Promotion of an institutional environment to sustain CDSP and similar interventions.
- Accumulation and dissemination of data and knowledge on the coastal areas.
- Direct improvement of the economic and social situation of people in a number of coastal chars areas in a sustainable way.

In its first phase, **CDSP-I** (1994-1999) developed three chars covering 6,800 ha in Noakhali District: Char Baggar Dona II, Char Majid and Char Bhatir Tek. A wide variety of activities were undertaken, ranging from infrastructure and water management to community development and health. Experience with char development in LRP and CDSP-I had led to the accumulation of considerable knowledge of the physical and socio-economic characteristics of the char areas as well as the potentials and constraints in char development.

An important factor that shaped **CDSP-II** (1999-2005) was the Integrated Coastal Zone Management (ICZM) concept that started to gain impetus in the late nineties (see Box 1). With the establishment of the ICZM framework, the demand for the experience gained in the coastal areas increased. During CDSP-II, there was room to pay proper attention to increasing the knowledge base in char development as well as to the dissemination of this knowledge.

[1] To develop the ICZM concept, a Project Development Office (PDO-ICZM), located under the Water Resources Planning Organization (WARPO) was established in 2002.

Box 1: The ICZM Concept

The specific coastal ecosystems, the vulnerability of the coastal zone due to natural disasters, and the land use conflicts justify a specific development approach for the coastal zone. The need for such a specific approach was already recognized in the early eighties. Initiatives from the Government of Bangladesh (GoB) and some donor agencies (UNDP, FAO) did not get the proper follow-up because of lack of political support at that time.

In the late nineties, the ICZM concept again gained momentum through initiatives of the Government and the donor community led by the World Bank. In 1999, the GoB produced an ICZM Concept Paper (Integrated Coastal Zone Management: Concepts and Issues; 1999) and a large-scale ICZM programme was planned to be launched in 2002; ongoing programmes and projects in the coastal zone, like CDSP, would become part of this ICZM programme. With the withdrawal of the World Bank from ICZM, the implementation of the ICZM concept lost part of its thrust. The Government of the Netherlands (GoN) and later on DFID continued with ICZM and funded the first step in the development of the ICZM concept; the elaboration of an ICZM policy framework.

It is the responsibility of the Ministry of Water Resources and the Water Resources Planning Organisation to take a lead and follow up the objectives of ICZM.

CDSP-I (1994-1999)
4 polders (CBD – I, CBD –II, Char Majid, Bhatir Tek) in Noakhali District
CDSP-II (1999-2005)
7 areas, including 5 non-poldered areas in Lakshmipur, Noakhali, Feni and Chittagong Districts.
CDSP-III (2005-2011)
2 areas in Noakhali and Lakshmipur Districts
CDSP-IV (2011-2018)
5 Chars (Char Nangulia, Noler Char, Caring Char, Urir Char and Char Ziauddin) Command area: 30,683 ha

CDSP-II contributed to the development of the ICZM framework through participation in the ICZM forums. The ICZM context in CDSP is reflected in the project objectives (see Table 2.3) and in the incorporation of the 'ICZM building blocks': (i) study on fresh water storage, (ii) study on coastal agriculture, (iii) the local level planning pilot. For the pilot on Local Level Planning, assistance was provided to four Unions in four Districts in participatory planning as well as in building the Union database. The resulting Union Development Plans were supported by partly financing some priority projects.

CDSP-II also covered a larger project area (33,000 ha with a population of 400,000), but with a more limited range of activities and a stronger institutional base. It took a more regionally based approach and dealt with both protected and unprotected areas.

CDSP-II resulted in a number of 'lessons learned'; there are two 'lessons' which are of particular relevance: (i) the sustainability of CDSP interventions and closely related to this (ii) the internalisation of char development concepts as developed during CDSP-I and II into the government agencies (see Chapter 13 for detail).

CDSP-II was followed by **CDSP-III** (2005-2011) which was meant to consolidate and monitor the achievements of earlier phases, while at the same time embarking on an intervention programme, specifically in Boyer Char covering about 6,500 ha in Noakhali and Lakshmipur districts. Also, the project needed to establish a bridge to a future char development programme by undertaking feasibility studies.

CDSP-III was followed by **CDSP-IV**, which was implemented from March 2011 until December 2018. The project was financially supported by the United Nations International Fund for Agricultural Development (IFAD), the Netherlands Government, and the Government of Bangladesh. The focus of the activities of CDSP-IV was on the development of five new chars: Char Nangulia, Noler Char and Caring Char (these three chars are contiguous); Urir Char and Char Ziauddin. The total extent of these chars is around 25,000 ha, with an estimated population of 170,000 in 29,000 households (See Table 2.2). The chars are located in Noakhali and part of Chittagong district. See the map of the CDSP-IV area in Figure 2.1.

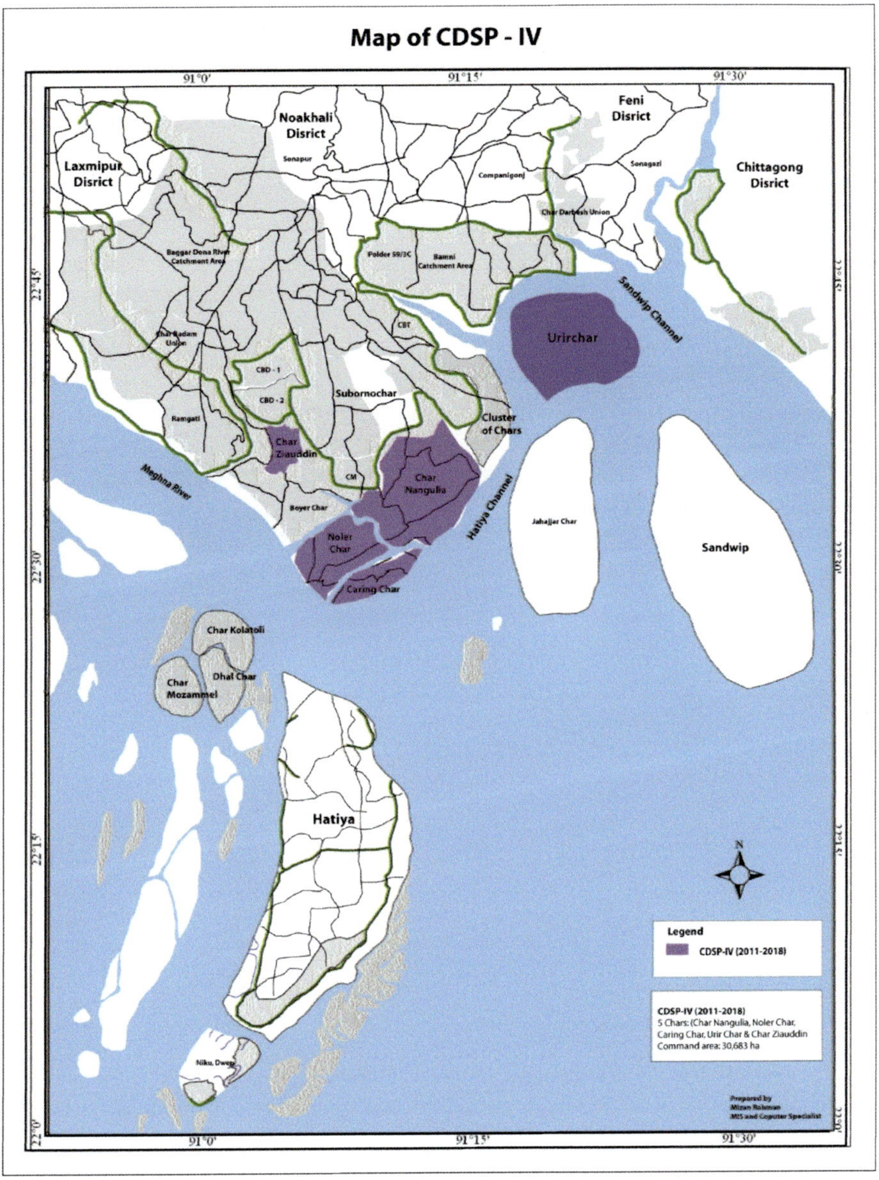

Fig. 2.1. Map of the CDSP-IV area.

Table 2.2. Area, population, and households (HH) of chars under CDSP-IV.

Name of the *char*	Area (hectare)	Population	Households
Char Nangulia	8,530	89,000	15,000
Noler Char	2,560	36,000	6,000
Caring Char	2,200	15,500	2,600
Urir Char	10,824	16,500	2,725
Char Ziauddin	1,943	14,000	2,380
Total	**26,057**	**171,000**	**28,705**

These chars accreted over 20-40 years and settlement started 11-15 years ago, except for Urir Char. In Urir Char, people were settled before the 1970s. The settlers are mainly from river eroded adjacent areas like Hatiya, Bhola, Ramgati, Companiganj and Sandwip.

The project continued support to the areas of CDSP-I, II and III with Operation and Maintenance activities and land settlement (in particular in Boyer Char). It also conducted feasibility studies in areas where future char development programmes might be undertaken.

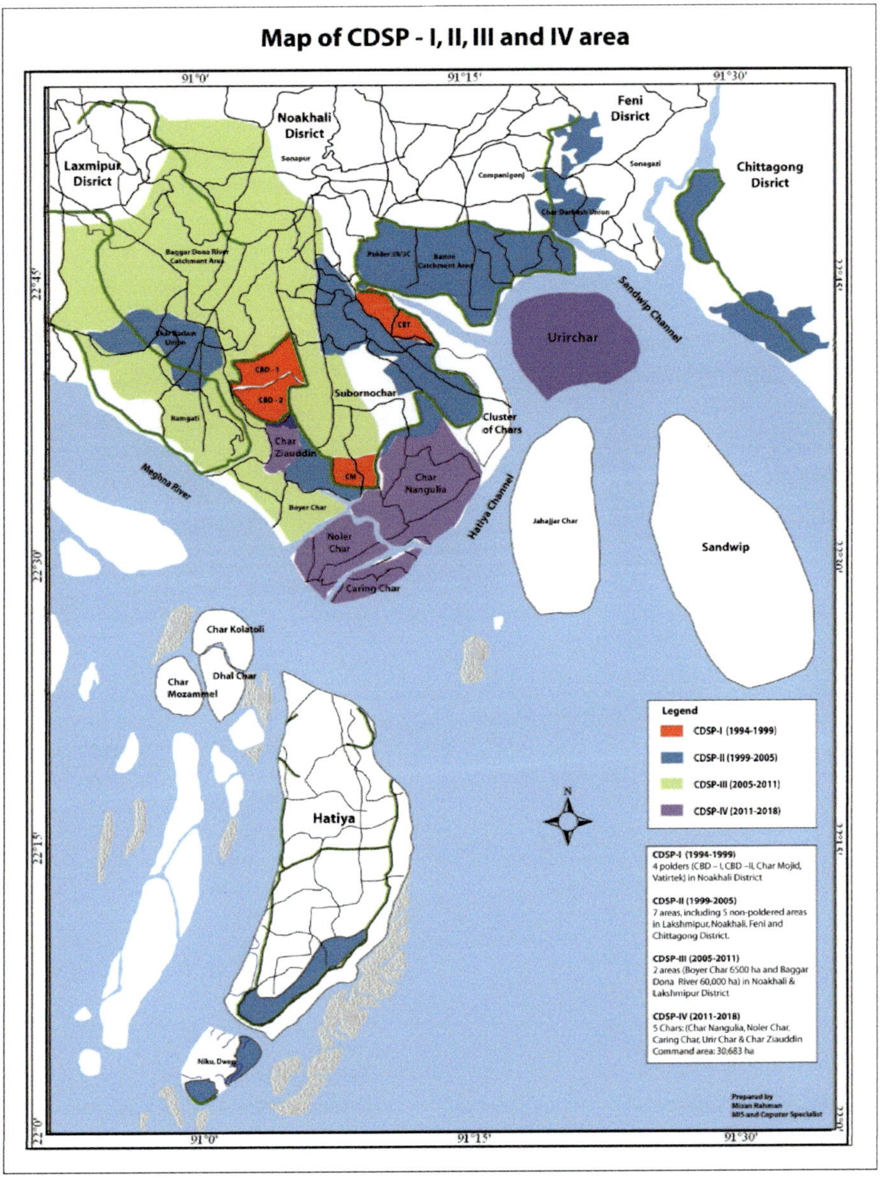

Fig 2.2: Map showing the location of the different phases of CDSP.

In CDSP-IV, there were six project components: protection from climate change; climate change resilient infrastructure, and water supply and sanitation; land settlement and titling; livelihood support; institutional development; knowledge management. The components and their direct linkage to project objectives are listed in Table 2.3.

The objectives at output level directly follow the results of the actual project activities and do indeed address the major problems that are at stake in the chars. In terms of substance, major challenges can, to a large extent, be addressed independently from each other. There are however interconnections at a practical level. For example, number and size of bridges and culverts (part of output "climate resilient infrastructure") have a direct bearing on the water management in the area concerned (output "water resources managed effectively"). Having a title on the land that a family occupies (output "secure land titles"), will substantially contribute to their socio-economic position and capability to cope with the vulnerabilities in the chars (output "improved livelihoods and household resilience"). In addition, all these four outputs feed, for a large part via the Monitoring and Evaluation system, into the output "knowledge management and lessons for ICZM", and vice versa.

Table 2.3. CDSP-IV components and the respective objectives.

Component	Sub-components	Activities	Objectives at Output levels
1. Protection from climate change:	1.1 Water management 1.2 Social forestry	(a) sea dykes; (b) internal embankments; (c) drains and canals; (d) water control sluices; (e) Water Management Organisations; (f) water infrastructure maintenance; (g) formation of social forestry groups; (h) tree planting on embankments, roadsides, foreshores, mudflats etc; (i) plantation caretaking	Effective management of water resources, protection against tidal and storm surges, improved drainage
2. Climate resilient infrastructure:	2.1 Internal infrastructure 2.2 Water and sanitation	(a) village and union roads and bridges; (b) cyclone shelters & killas (animal shelters); (c) rural markets; (d) deep tube wells; (e) drinking water ponds and rainwater collection; (f)	Climate resilient internal infrastructure for communication, markets, cyclone shelters, provision of potable water and hygienic sanitation

Component	Sub-components	Activities	Objectives at Output levels
		hygienic latrines; (g) Labour Contracting Societies for construction; (h) O&M user groups; (i) market management committees; (j) infrastructure maintenance	
3. Land settlement and titling:		(a) surveys to assess availability of land and current ownership status; (b) selection of target group households; (c) process of land titling; (d) computerized land record management system	Provision to the settlers of a legal title to land
4. Livelihood support:	4.1 Agricultural support 4.2 Social and livelihood support	(a) formation of groups; (b) identification of appropriate technologies; (c) capacity building of service providers; (d) crop training and demonstrations; (e) other skill training; (f) access to livelihood opportunities and markets; (g) promotion of better health and hygiene; (h) social support and rights; (i) disaster preparedness and climate change resilience	Improved livelihoods and household resilience
5. Institutional development:		(a) field level institutions (FLI); (b) local	Institutional development in order to create

Component	Sub-components	Activities	Objectives at Output levels
		government institutions; (c) Gender Action Plan (GAP)	an enabling institutional environment
6. Knowledge management:		(a) studies and surveys; (b) Monitoring & Evaluation	Knowledge management through undertaking and disseminating surveys and studies and by learning from and contributing to ICZM efforts

Robert Chambers' fundamental concept of five livelihood capitals, namely: human capital, social capital, physical capital, natural capital and financial capital, form the pillars of the sustainable livelihood development approach and have particular relevance to the CDSP interventions, as they aim to strengthen all of these five capitals. See below for the five capitals and the interventions of CDSP-IV for each of these capitals.

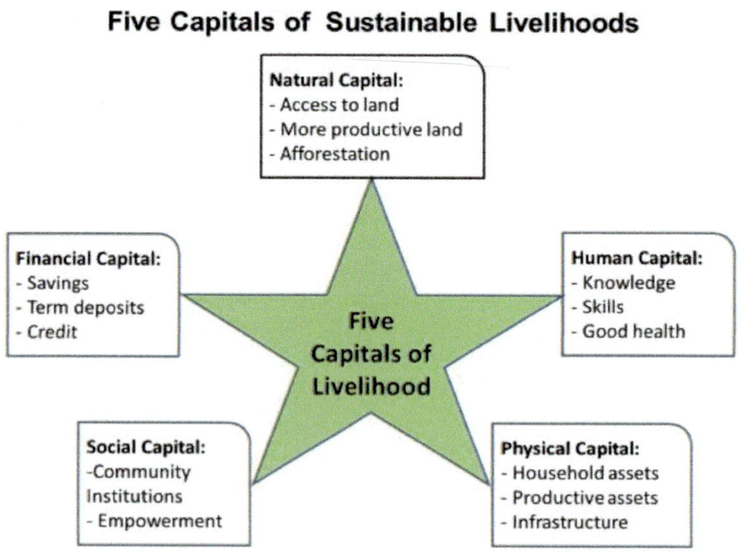

The five different livelihood capitals interact with each other, with each one supporting and enhancing the other four capitals. This is shown in **Table** 2.4. The horizontal rows show how each of the capitals named in the left-hand column enhance and support each of the capitals named across the top row of the table. These five capitals have combined to increase household income significantly and enable households to diversify their livelihoods into new farm and non-farm enterprises (CDSP-IV TR 13, 2017).

"The people of the coastal area have won land, agricultural fields and necessary infrastructure. We hope that soon the chars will no longer be poverty hotspots", Bazlul Karim, deputy team leader of CDSP-IV.

Table 2.4. Complementary linkages between livelihood capitals.

	Natural	Human	Physical	Social	Financial
Natural			Income from farming land used to purchase assets	Obtaining a secure title to land is empowering, especially for women	Formal ownership of land makes household more credit-worthy
Human	Improved knowledge and skills help make land more productive		Knowledge and skills are needed to use livestock and productive assets	Improved knowledge is socially empowering	Reduced danger from ill health reduces risk in taking micro-loans
Physical	Water control infrastruc-ture makes land more productive	Schools in cyclone shelters educate children. Roads improve access to health services.		Roads enable social connectivity. WMG centre buildings strengthen these institutions	Roads make micro-finance provision easier. Water control infrastructure means loans are less risky.
Social	FLI monitor land allocation and titling process	FF dissemin-ate know-ledge on farming	WMG operate water control infrastructure. TUG		NGO groups operate micro-finance services

	Natural	Human	Physical	Social	Financial
			maintain DTW		
Finan-cial	Loans invested to lease in and improve productiv-ity of land	Loans also used for education and health expenses	Loans and savings invested in assets and livestock	Social awareness raising at micro-finance meetings	

The positive changes achieved through the project in the five livelihood capitals are discussed in detail in Chapter 11. CDSP is an example of integrated coastal zone management (ICZM) in Bangladesh. The principle of ICZM is that development problems and constraints needed to be tackled by concerted, rather than isolated interventions as poor people are not able to compensate for the missing links. The strong focus on the integration of the different approaches in addressing the multi-dimensional aspects of poverty reduction and food security makes this project stand out among the many single-sector projects dealing with poverty reduction and food security elsewhere in the country. The integration approach is based on the premise that the social and economic situation in the chars, defined by a set of vulnerabilities, cannot be meaningfully improved by one single intervention, nor by one government agency. CDSP is a multi-discipline and multi-agency development effort, undertaking a set of different interventions within the same geographical area in the timeframe of a project. The administrative basis is an umbrella Development Project Proforma (DPP), with separate DPPs for the participating agencies. This common planning and coordinated implementation, but with each agency doing what it is best at and each agency with its own money flow, is now recognized by many as being a "best practice" method of implementing multi-sectoral programmes. CDSP-IV was implemented by six government agencies with the Bangladesh Water Development Board (BWDB) as the lead agency. The roles and responsibilities of these Government of Bangladesh (GOB) agencies have been as follows:

i) **Bangladesh Water Development Board (BWDB)**: construction and maintenance of all water management related infrastructure such as embankments, sea dykes, sluices, irrigation inlets, drainage khals, closure of khals and culverts.
ii) **Local Government Engineering Department (LGED):** construction of all non-water internal infrastructure, such as rural roads, bridges and culverts, cyclone shelters, and houses.

iii) **Department of Public Health Engineering (DPHE)**: responsible for public water supply and sanitation, with installation of deep tube wells and latrines.
iv) **Department of Agricultural Extension (DAE):** responsible for all activities related to agricultural development, in particular field crops.
v) **Ministry of Land (MOL):** responsible for all activities related to the process of land settlement and for the strengthening of the land settlement bureaucracy.
vi) **Forest Department (FD):** responsible for foreshore mangrove and non-mangrove plantations, embankment and roadside plantations, nursery development and community mobilisation with training.

The project's Technical Assistance (TA) team ensures the integration and coordination of the activities of the different ministries and departments of the Bangladesh Government in the project area. The integration in CDSP went further than only the government agencies. In the coordination mechanisms NGOs were also represented by an NGO programme which was implemented in the same area and in the same period.

Table 2.5. Parties involved in CDSP-IV and their respective responsibilities.

	Agency	Responsibility
A.	**Government Agencies**	
	Bangladesh Water Development Board (BWDB)	Delegated Lead Agency and responsible for main protective and water management infrastructure;
	Department of Agricultural Extension (DAE)	Agricultural extension
	Department of Public Health & Engineering (DPHE)	Water supply and sanitation
	Forestry Department (FD) (since 2005)	Forests and afforestation
	Local Government Engineering Bureau (LGED)	Internal infrastructure
	Ministry of Land (MoL)	Land settlement
B.	**NGOs**	
	Sagarika Samaj Unnayan Sangstha (SSUS)	Responsible for specific chars
	Dwip Unnayan Sangstha (DUS)	
	BRAC	
	Society for Development Initiatives (SDI)	
C.	**Door Agencies**	
	International Fund for Agricultural Development (IFAD)	Financing, monitoring
	Embassy of the Kingdom of the Netherlands (EKN)	Financing, monitoring
D.	**Technical Assistance (TA)**	Advisory and monitoring

Climate Change

Harsh natural conditions (i.e. flooding, cyclones and a lack of fresh water) and an unfavourable socio-economic situation (i.e. lack of a secure land title, poor communication infrastructure, lack of institutions and governance) in char areas mutually reinforce chronic vulnerability to climate conditions and natural disasters. Although people have always adapted to the climate, the current process of climate change coupled with extreme vulnerability and exposure makes increasing the adaptive capacity of the char population essential, especially for the medium and long term.

As it is impossible to anticipate exact future impacts of climate change – particularly at local scale – climate change adaptation should have an intermediate goal of empowering communities to adapt to the impacts in a broader development perspective, which is exactly what CDSP-IV is doing. CDSP-IV includes both community-based and engineering interventions, which has proved to be a very effective approach as it not only addresses climate change adaptation but also increases the resilience of communities to current climate conditions, while simultaneously improving the socio-economic situation in the chars.

The overall objective of CDSP-IV is to improve the economic situation and living condition of the population in the coastal areas of South-eastern Bangladesh with special reference to the poorest segment of the population. In doing this the project reduces current vulnerability to climatic conditions and increases the capacity of the char communities to adapt to climate change impacts. Along with providing physical security through infrastructural development, CDSP has also assisted the char dwellers in receiving legal land titles which is the first priority of the settlers.

Integrated Development

Experience from CDSP I, II, III and IV shows that, thanks to the reduced vulnerability provided by the infrastructure, combined with the security of land tenure, higher incomes, and other socio-economic improvements, people have invested in better housing, more resilient agriculture and effective local institutions capable of dealing with changing conditions. This holistic approach in CDSP with interventions across a wide range of different sectors is a model for the development of coastal communities to avoid the future risk of climate change. The Government of Bangladesh is now trying to replicate this model in other char areas to help the coastal communities to adapt to the changing climate.

Riverbank Erosion

A major challenge for coastal development in Bangladesh is the continuous river bank erosion. The major rivers in Bangladesh carry large amounts of sediments, especially during the monsoon season which makes the rivers braided and creates a migrating pattern of channels and char lands (BDP2100, 2017). It is very difficult to predict the accretion and erosion process in the Bay of Bengal. Prediction is only sometimes possible at medium term (10-20 years) to identify roughly which areas will likely accrete and which will erode. River studies are necessary on land erosion and accretion for the planning of any new project in char areas. For physical char development, two other aspects are also important: the minimum land level for empoldering and the accretion rate in front of planned polders in view of drainage opportunities.

Fig. 3.1. Riverbank erosion in Shantipur, Noler Char, 2018.

Project coordination and internalisation

CDSP-IV is unusual in development terms, in that it successfully brings together ten different organisations. The programme is co-financed by the Government of Bangladesh, the Government of the Netherlands, and IFAD, and relies on the cooperation and cross-sectoral information sharing of six government departments, namely: Bangladesh Water Development Board, Forest Department, Local Government Engineering Department, Department of Public Health Engineering, Department of Agricultural Extension, and the Ministry of Land. Four partner Non-Government Organisations are also taking part: Sagarika Samaj Unnayan Sangstha, Dwip Unnayan Songstha, Society for Development Initiatives and BRAC. The large number of parties involved with the project requires appropriate coordination mechanisms at the national and local levels. For example, in order to avoid a potential tension between the advisory and support functions with regard to the implementing agencies on the one hand, and the monitoring of quality control and recommendations on payments on the other hand, meetings between the agency concerned and management of the Technical Assistance (TA) team are organized to facilitate a situation in which both functions can be fulfilled successfully.

The activities for which the TA team has an initiating and leading role always take place under the responsibility of the Project Management Committee and the Project Coordinating Director of the BWDB. In addition, the sub-contracting, and in most cases tendering, for those activities for which that is required (for instance NGO selection, commissioning of studies) are done by the main consulting firm.

Consolidation of achievements of earlier phases, is one of the aims of CDSP. Internalisation is understood as incorporating concepts and experiences of CDSP I, II and III into the approaches and working methods of the participating government agencies. The aim is to make these agencies better prepared for future char development programmes. This has proven to be a difficult task, with limited but, at the same time, significant achievements.

Alignment with Country Policies

CDSP is very well aligned with MDGs, SDGs, and several important government policies which are mentioned in the following section.

National Poverty Reduction Strategy (National Strategy for Accelerated Poverty Reduction 2009-11)

The second poverty reduction strategy identified chars as being a pocket of extreme poverty and it specifically mentions the Char Development and Settlement Project – which is designed to support economic development and poverty reduction. In this strategy, water management, agriculture, forestry, rural roads, land policy and disaster management are all focal areas for pro-poor growth, to which the project is contributing. Supporting strategies include actions to reach extreme poor groups, support for better water and sanitation, especially where groundwater conditions are unfavourable (such as the saline coastal area), and adapting to climate change (CDSP-IV, 2012).

The Coastal Development Strategy (2006)

CDS has evolved around nine strategic priorities:

- Ensuring fresh and safe water availability
- Safety from man-made and natural hazards
- Optimising the use of coastal land
- Promoting economic growth emphasising non-farm rural employment
- Sustainable management of natural resources
- Improving livelihood conditions of the people, especially of women
- Environmental conservation
- Empowerment through generating and disseminating information and knowledge
- Creating an enabling institutional environment.

CDSP arguably has strong credentials as far as seven of these strategic priorities are concerned, with relatively less emphasis on non-farm employment and on environmental conservation. The Coastal Development Strategy is very much formulated along the lines of the principles of Integrated Coastal Zone Development, and CDSP is at heart an ICZM-project, contributing in a significant way to applying ICZM in Bangladesh. In the eyes of the staff of the ICZM-project that ceased to exist in 2006, CDSP was an ideal environment for piloting certain ideas, such as establishing a District Information Centre as a dissemination vehicle for coastal information and creating a connection between data bases of coastal projects with the central coastal data base at the Water Resources Planning Organization (WARPO). These activities were indeed included in CDSP III. However, since 2006, implementation of the strategy has not taken place due to reduced funding and changes in governmental priorities.

Bangladesh Delta Plan 2100

Owing to the deltaic formation of the country, the configuration of the rivers and the challenges posed by natural disasters and climate change, Bangladesh has been ranked as the 5th most vulnerable country in the world in terms of risks from natural hazards. In view of the special long-term challenges for development outcomes presented by climate change and natural hazards, the Government has decided to formulate a long-term Bangladesh Delta Plan 2100 (BDP2100) with the help of the Netherlands Government. The BDP2100 looks primarily at the medium-term measures (2030-50) but also seeks to address the longer-term challenge of sustainable management of water, ecology, environment and land resources in the context of their interaction with natural disasters and climate change.

CDSP aligns well with these specific goals and covers 6 blocks among the 8 blocks of integrated issues covered. Among six components of CDSP-IV, four of them directly and indirectly address the impacts of climate change. This is done through, for example, effective management of water resources for protection from climate change, constructing climate resilient infrastructure, and livelihood support to improve livelihoods and household resilience. In addition, the 'Institutional Development' and 'Knowledge Management' component of CDSP-IV covers the last block of the Delta Plan, which is Governance.

BDP2100 specific goals

Goal 1: Ensure safety from floods and climate change related disasters;
Goal 2: Enhance water security and efficiency of water usages;
Goal 3: Ensure sustainable and integrated river systems and estuaries management;
Goal 4: Conserve and preserve wetlands and ecosystems and promote their wise use;
Goal 5: Develop effective institutions and equitable governance for in-country and trans-boundary water resources management;
Goal 6: Achieve optimal and integrated use of land and water resources.

Agricultural Khas Land Management and Settlement Policy 1997 (including 1998 Amendment)

State-owned land is known as khas land in Bangladesh. In 1997, the Khas Land Settlement Policy was promulgated, authorising khas land to be distributed for a 99 year lease period. CDSP works in 'char land' which is state-owned land and falls into the broader category of khas land. The aim of the project is to improve the security of landless people who are informally occupying char land by providing them with legal title. The land settlement process of CDSP follows the provisions of agricultural khas land management and settlement policy of 1997. Over the years of CDSP implementation the process has been streamlined and simplified making it faster and more accessible for the char settlers.

Social Forestry Rules 2004 (Amended in 2010 and 2011)

In Bangladesh, involvement of local people in forest management has changed much in recent years. Since the 1980s, forest people came to be viewed as important stakeholders for forest conservation rather than destroyers of the forest. Forest law has been changed to facilitate social forestry and to strengthen forest protection by involving local people. The Forest (Amendment) Act 2000 facilitates social forestry in Bangladesh and gives the concept of social forestry a clear shape. In 2004, Social Forestry (SF) Rules came into force which state that, for efficient management of SF a nine-member SF Management Committee for each locality of social forestry is constituted. The members of the committee are elected by the beneficiaries of the concerned social forestry locality with a minimum of one-third being elected from women participants (Jashimuddin, 2012).

CDSP-IV is implementing different types of social forestry programmes and to maintain these plantations, the project uses Social Forestry Groups (SFGs) who are the core of the social forestry approach.

Guidelines for Participatory Water Management (2001)

In 2001, the Ministry of Water Resources (MoWR) shifted responsibilities for water management away from state implementing agencies towards community-based Water Management Organisations (WMOs), with limited involvement of local government institutions, by formulating the Guidelines for Participatory Water Management (GPWM) in 2001 within the framework of National Water Policy, 1999 (Dewan et al, 2014).

The MoWR defined Participatory Water Management (PWM) Rules through a gazette notification in 2014. The Rules relate to formation and functions of WMOs in water resources projects. The most important shift which the PWM Rules 2014 suggest is that from now on the WMOs will be registered under Bangladesh Water Development Board (BWDB) rather than the Department of Cooperatives (DoC). The Rules emphasise a larger enrolment of local stakeholders in the Water Management Groups (WMGs) and their participation in Operation and Maintenance (O&M) of the water management infrastructure.

CDSP follows the GPWM to form WMOs in project areas for the development of the institutional framework and capacity of the local stakeholders to gradually establish ownership through participatory water management.

Some of the lessons learnt of CDSP regarding managing this complex process of char development are:

- Every char is different in terms of their dynamic environment (i.e. sea facing or hinterland) and socio-economic situation. Thus, requires a flexible planning with maximum time horizon of 20 years. This is however challenging for projects as activities need to fit within a project time horizon of 5 to 7 years.
- As mentioned above in this chapter a major challenge for coastal development in Bangladesh is the continuous river erosion making it difficult to predict the accretion and erosion process in the Bay of Bengal. Prediction is only sometimes possible at medium term (10-20 years) to identify roughly which areas will likely accrete and which will erode.
- Char development is an integrated process and cannot be done in isolation. We have to consider the surrounding environment during planning such as future accretion of the surrounding areas to determine the drainage design and O&M. This is an important issue to consider during a feasibility study.
- Char development should be embedded in wider government plan. CDSP is mentioned in Bangladesh government planning documents and the ICZMP as the way to develop new chars. However, account needs to be taken of competing uses of new chars such as fish/shrimp farming, and Rohinga resettlement.

Chapter 4

Involving the Communities and Civil Society

Md. Bazlul Karim

It has been demonstrated in many parts of the world that development initiatives are more effective for poverty reduction when all stakeholders, especially citizens and marginalized communities are actively involved in the planning, implementation and monitoring of development programmes. Moreover, effective development calls for the "ownership" of processes of change by those who will embody them in the future. In most cases, external interventions have negative implications at local level because they often lead to a lack of ownership on the part of local communities. To avoid such a situation, CDSP has adopted a multi-dimensional approach which requires an institutional basis at community level in order to promote a participatory process as well as to make the efforts sustainable (CDSP-IV PR 14, 2018). Since the beginning, CDSP has been involving the settlers with planning and implementation of project activities. In addition, from early on in the process (during the feasibility study phase) local government institutions are involved in the project to ensure their support and assistance in implementing different project activities. The following section describes how CDSP is involving and empowering the char communities in different ways to continue the development of their area.

Local Government Institutions

Union Parishad (UP) is the smallest of the rural administrative and Local Government Institutions (LGIs) in Bangladesh. Each UP consists of a chairman, nine general members, and three women members. UPs provide assistance through their representation as advisors to the concerned Water Management Organizations (WMOs) in respect of Participatory Water Management (PWM) at the local level. This is done through the respective standing committees. These roles are specified in CDSP as below:

a) UP provides assistance through representation as advisor of the concerned WMOs and FOs, and also through their respective standing committees.
b) UP acts as a coordinating agency for linkage between WMO and the public-sector agencies.
c) UP ensures the security of the water management infrastructure.
d) WMOs are responsible for taking part in Operation and Maintenance (O&M) of medium hydraulic structures, replacement of fallboards, repair of minor flood damage work, medium erosion protection and painting gates of medium sized structures. UP provides supporting, facilitating and coordinating assistance to the WMOs in respect to PWM.
e) UP provides institutional support to the WMOs, specifically to arrange volunteer labour inputs and financial contributions from beneficiaries on O&M.
f) UP provides support to the WMOs for awareness raising and participation of people in the water management groups.
g) UP monitors the activities of WMOs.
h) UP representatives participate in the WMO meetings.
i) UP provides support to WMOs with respect to raising funds for O&M.

A process has started to formalize the participation of LGI representatives in particular in Water Management Association and Farmers' Association meetings and subsequently WMO and FO representatives' participation in UP coordination meetings to deal with local issues. Specifically, water management issues require wider participation to be resolved jointly and to promote a process for developing partnerships at local level.

Non-Government Organizations

In every phase of CDSP there have been several NGOs involved as development partners for livelihood and social development activities. In the current phase, four NGOs are involved with CDSP-IV interventions and work simultaneously in several components to implement multidimensional programme interventions. The focus is on social and economic development of households in the five chars, especially women and children. The selected partner NGOs (PNGOs) and their working areas are shown in Table 4.1.

Table 4.1. PNGOs' assigned areas.

No.	Name of NGO	Name of working chars
1	BRAC	Char Nangulia, Char Zia
2	Sagorika Samaj Unnayan Sangstha (SSUS)	Char Nangulia, Noler Char
3	Dwip Unnayan Sangstha (DUS)	Noler Char
4	Society for Development Initiatives (SDI)	Urir Char

Each PNGO has a programme coordinator to supervize its activities in the various branches and the TA team provides both technical and management support. A brief discussion of the PNGO activities is provided below.

Health and family planning

Being located in remote areas, char dwellers suffer from lack of proper and timely healthcare facilities. To address this problem, the project provided medical services like health and nutrition education, safe water and sanitation education, family planning, immunization, pregnancy related care and safe delivery and some basic curative facilities. These services were provided in the char areas through paramedics, shasthya shebikas (female health workers), and trained traditional birth attendants (TBA) (BRAC, 2011).

Micro-credit and capacity development

Access to credit is a key requirement for improved livelihoods and poverty reduction. When infrastructure development in the CDSP I, II and III areas was complete, Micro Finance Institutions (MFIs) set up their operations and offered micro-finance services. However, it was realised in CDSP-IV that there is an immediate need for micro-finance that cannot wait until infrastructure is developed. Now, MFIs set up their operations mainly to encourage group savings and to generate income that makes people less dependent on donor funding and allows them to sustain operations after the end of the project. In addition, micro-finance groups meet regularly which is a good forum for providing technical training and social support. New technologies can be promoted through a combination of training and loans for investment.

Case profile of Zohura Khatun (Improved TBA service)

Zohura Khatun, aged 38 years, has been working in Hazi Idris Miah Bazar of Char Nangulia for a long time as a TBA. However, she has never received any formal training for this work. Before the CDSP-IV intervention, she occasionally faced difficult situations and didn't know what to do. Consequently, many vulnerable pregnant women died due to the lack of proper advice or diagnosis of the symptoms. Zohura was also not aware of the importance of colostrum (also known as first milk) which contains antibodies to protect the new-born against disease. Instead, she advised the lactating mothers to feed their babies 'honey' or some other liquid. As a result, most of the babies developed stomach sicknesses. Recently, Zohura got involved with CDSP-IV where a 15-day TBA orientation was organized by the health and family planning staff of the project. The TBA orientation programme covered all the critical situations related to delivery, and pre and anti-natal care. Since receiving the training, she has observed the dramatic qualitative change in her work. Nowadays, Zohura can identify the signs of a vulnerable and critical situation during pregnancy. Whenever she identifies a risky situation, Zohura advises them to go to the local hospital where a safe delivery can be performed with proper medical support for both the baby and the mother. She also now advises the new mothers to feed the first course of breast-milk (colostrum). Zohura thinks babies and mothers are much healthier nowadays than before the intervention of CDSP-IV. Thus, she is grateful to the project for the orientation she received for TBA work.

Case profile of Nahar Begum (Training in tailoring passed on to a young woman)

Nahar Begum, aged 19, lives in Char Ziauddin. She is the third child of five siblings. Her family used to live in Bhola but unfortunately, due to four episodes of erosion by the Meghna river, they lost everything, including their house and cultivable land. Through local people Nahar's father came to know that a new char had emerged in the river, called Char Ziauddin. Nahar's father migrated there with his full family in 2001. He took possession of an area of 0.4 ha of land in exchange for paying Tk.3,000 to a samaj (community) leader. After clearing the forest, Nahar's family started living in Char Ziauddin. At that time Nahar was a two-year-old child. Nahar had studied up to class five in Ziauddin char when, in 2014, aged only 15, she got married to Ruhul. However, Nahar returned to her parents after only six months of marriage as she could not cope with her husband. As she had left her husband, neither her husband's family nor her community supported this decision. Nahar's mother was, by then, a member of a CDSP-IV group, and heard that training in tailoring would be provided to women by this project. However, the training was limited to group members only, so Nahar did not qualify to receive this training. However, she was told that a woman from the neighbouring community had received the training. Nahar then asked the woman to train her, which the woman did, providing Nahar with tuition for three months. Now, she has become well-known as a tailor in her area. In November and

December, she earned Tk. 4,000 by making school uniforms for girls. She also earned Tk.2,000 by providing two girls with training in tailoring. Nahar now wants to become a trainer of tailoring for girls in an organization. Along with that, she will also continue her tailoring activities from her home.

Climate change and disaster management

Drought, excessive rainfall, floods, waterlogging, etc. are some of the most common disasters faced in the coastal chars. CDSP attempted to raise awareness about these disasters and ways to cope with them among both the members of the groups and the community in general (BRAC, 2011). During the regular group meetings, the partner NGOs arranged discussions on pre and post-disaster preparedness, the significance of different warning signals, communication and linkage with Red Crescent and local responsible people, locations of cyclone/emergency shelters, ways to protect assets, facilitating timely and effective rescue, relief and rehabilitation, community-based preparedness and management and so forth.

There is a Union Disaster Management Committee (UDMC) in every Union Parishad. To strengthen and involve UDMC, meetings are organized by PNGO's Disaster Coordinators with UDMC and other (e.g. Red Crescent) stakeholders where various issues related to disaster management and options for better cooperation among the stakeholders are discussed.

Water and sanitation

Char areas often lack safe drinking water and hygienic sanitation facilities. Before the intervention of CDSP, latrine usage by the community was very poor, on average less than 4%, resulting in diarrhoeal diseases. The project aimed to ensure easy access to safe water supply and hygienic sanitation facilities among the char dwellers by providing Deep Tube Wells (DTW) and sanitary latrines from the Department of Public Health Engineering (DPHE) (see Chapter 6) under the supervision and coordination of CDSP-IV. The PNGOs supported the DPHE by assisting in: the formation of tube well user groups (TUG) comprising one woman member per family, site selection for DTW installation, collection and depositing of contribution money to DPHE and ensuring capacity building support to the beneficiaries.

Legal and Human Rights (LHR)

CDSP-IV aims to make char dwellers, especially deprived women, aware of human rights and basic laws through better access to information. Volunteers received a one month training on this topic. Once trained, they organize meetings among the NGO group members to make people aware of legal and human rights and social issues. In addition, 20% of the beneficiaries received training on legal and human rights on seven basic laws. These are: 1) Muslim family marriage law, 2) Hindu family marriage law, 3) Muslim inheritance Law, 4) Hindu inheritance law, 5) Land law, 6) Criminal law, and 7) Bangladesh constitution law. These 20% of beneficiaries are members of the group management committee and act as a law implementation committee, disseminating the message and acting against any violation of human rights.

Homestead agriculture and value chain development

This programme aimed to reduce malnutrition and increase the income of households by engaging women in economic activities within their homestead. This involved technology transfer, training, demonstration plots, and other capacity building activities. The programme objectives are:

- to utilize maximum resources of the homestead for growing vegetables and fruit;
- to promote new technologies around homesteads;
- to enhance intake of vegetables and fruit to ensure family nutrition;
- to increase cash income and facilitate women's empowerment; and
- to contribute in improving the livelihood condition of char dwellers.

There is also a Value Chain Development (VCD) programme to increase the income of the farmers through linking them to different market actors. There is a value chain coordination committee headed by Project Director CDSP-IV (DAE part), PNGO representatives, the TA team to facilitate and ensure proper implementation of the value chain programme. In the project area, 520 growers and 120 market actors were selected for a comprehensive orientation and training on value chain development. After the training, the market actors and growers selected several products for value chain development. The selection was done through participatory methods considering market demand, growth potential and potential number of people engaged in production. Once the value chains are selected, the groups carry out in-depth analysis of different channels/segments, constraints, opportunities and remedial measures. Producers are linked with the buyers, suppliers, local technology providers and the government agricultural extension office which provided them with technology training. Some of the priority products selected for value chain development in CDSP-IV are: okra, cucumber, country beans, country bean seeds, tomatoes, bitter gourd, sweet gourd, soybeans, green chillies, watermelons, methi and local bananas.

Fisheries

Bangladesh is one of the richest countries in the world in respect of water resources. The country has vast water bodies in the form of rivers, canals, estuaries, haor, beels, lakes and ponds. Fisheries play an important role in nutrition, employment generation, poverty alleviation and foreign exchange earnings.

Up until 2012, the Regional Fisheries and Livestock Development Component (RFLDC), a DANIDA funded project, was implemented in CDSP areas. In order to avoid duplication, fisheries and livestock development activities were not included in CDSP-IV. However, after the phasing out of RFLDC, project beneficiaries reported a lack of services from the government or non-government side and requested such assistance from CDSP-IV. Thus, fishery activities have been included in CDSP-IV since 2014 to improve the livelihood and socio-economic status of char dwellers by fish farming. Some of the objectives of the fisheries programme are to:

- cover maximum water bodies under improved fish culture;
- develop the knowledge of fish farmers and promote modern and sustainable technology by providing training;
- enhance protein consumption;
- ensure the availability of quality fish fry for fish farmers;
- reduce fish diseases in the project area; and
- ensure the overall sustainability of fish culture.

Case profile of Kuhinur Begum (The path from poverty to prosperity)

Kuhinur Begum, aged 35, lives in Char Ziauddin with her husband, one daughter and two sons. She migrated to Char Ziauddin in 2002 from Ramgati upazila of Noakhali district due to river erosion. Kuhinur's husband was a day labourer with very little income. In 2012 BRAC (PNGO of CDSP-IV) started its activities in Char Ziauddin. Kuhinur got herself involved with a cooperative run by BRAC in CDSP-IV and started depositing Tk. 10 every week in the cooperative. Along with support such as sanitary latrines, training in vegetable cultivation, vermi compost preparation and employment through the Labour Contracting Society (LCS), she received 2 days training on fish farming as a model farmer in 2013. Kuhinur received her first loan of Tk. 8,000 from BRAC in weekly instalments for fish culture. Kuhinur and her husband dug a pond with the help of the loan money and their own capital and started cultivating fish. At the beginning, Kuhinur failed to make a profit as they had little knowledge about proper fish cultivation. After receiving 2 days training on "Improved Fish Culture Management" as a model farmer in 2016, she established a fish farm in her 18-decimal pond with the help of Tk. 25,000 from a Partner NGO of CDSP-IV. With this money, she released 1,000 different fingerlings (rui, katla, mrigal, silver carp etc.) in her fish farm as an input. She also received other technical support from the project. Kuhinur's production cost was Tk. 15,000 (dyke repair, fingerlings, lime, fertilizer and feed etc.). They produced 375 kg fish and sold the fish to local traders. She earned a total of Tk. 33,750 (Tk. 105/kg) from her farm after meeting her family needs. Kuhinur has expressed her interest in expanding her fish farm in future.

Fig. 4.1. Kuhinur with her husband collecting their fish catch.

Poultry and livestock programme

Poultry and livestock rearing is one of the important sources of livelihood for poor people in the rural areas of Bangladesh. Rearing of poultry (chickens, ducks and pigeons) and livestock (goats, cattle, and buffalo) is an important source of household economy and food, as well as social security. According to the 2014 baseline survey, the average number of birds for each household was 10 chickens and 5 ducks. However, rearing poultry and livestock using traditional methods frequently resulted in high mortality due to the lack of vaccination services. The poultry and livestock programme of CDSP worked to promote proper management and care of poultry and livestock among the participants, by providing proper training, inputs, and marketing services.

Field Level Institutions (FLIs)

Community-owned processes will only be transformative if they are driven and self-organized by internal forces and dynamics of change. In order to promote such a process in the CDSP areas, six implementing agencies and four PNGOs, with support from the Technical Assistance team, are working to strengthen capacity development of Field Level Institutions (FLIs). FLIs are established to involve local people in planning and implementation of project interventions such as infrastructure development for better water management, social forestry, agricultural extension and livelihood and in operation and maintenance after the project is completed (CDSP-IV PR 14, 2018).

These groups give shape to the concept of people's participation, an important element in the project's approach. The institutions formed in CDSP-IV are:

- Water Management Organizations (WMOs) to manage water control infrastructure. These WMOs operate at three levels: Water Management Groups (WMG) represent farmers within a geographically defined water management area; at polder level (i.e. one char with a surrounding embankment), a number of WMGs form a Water Management Association (WMA), with a Water Management Federation (WMF) at the district level. WMGs are registered as cooperatives by the BWDB in line with the rules promulgated by the Ministry of Water Resources.
- NGOs established groups to receive micro-credit and other services from NGOs, as well as Tube Well User Groups for proper maintenance of tube wells.
- DAE has established Farmers' Organizations (FO) as the focal point for the agricultural development activities. These FOs operate at three levels: Farmers Forum (FF) representing farmers within a village; at polder level (i.e. one char with a surrounding embankment) a number of FFs form a Farmers' Association (FA), with a Farmer's Federation (FF) at the district level.
- Social Forestry Groups (SFGs) are formed by the Forest Department to establish and maintain different types of tree plantation. The SFG members receive a percentage of profit gained from the social forestry activities (see Chapter 10).
- BWDB, LGED and DPHE form Labour Contracting Societies (LCS[2]) to undertake labour-intensive construction and maintenance works.

---

[2] A Labour Contracting Society (LCS) is an organized group of men and/or women who are contracted to carry out works for a government agency or project. Under public procurement rules they can be selected to implement works without going through a competitive procurement process – with the objective of providing employment for poor local people living in the vicinity of the works being carried out.

Membership of FLIs has empowered char dwellers to raise their voices in demanding their rights, increasing negotiation and bargaining power, and ensuring the participation of women in all spheres of life. Some of the benefits of joining FLIs mentioned by project beneficiaries in a recent impact assessment survey (2018) are recorded Table 4.2.

The FLIs established by CDSP-IV are not the end result of this endeavour; rather these are considered as effective means to promote a sustainable development process in the local areas. CDSP interventions will only be sustainable if the settlers actively participate in the process of planning and implementation of development activities. In addition, while forming these field level institutions, female participation is promoted. It is demonstrated that these institutions are instrumental in providing government services in remote areas where these services were not visible in the past. FLIs are now in the process of developing linkages with other market/social forces including their respective local government institutions as they have realised that they cannot address all the problems by themselves. Coordination among field level institutions is done by undertaking bilateral meetings among these institutions. The Water Management Groups (WMGs) coordinate with the Union Parishad and with LCSs through meetings. The WMGs have meetings with the UPs concerned and also, if required, meet the representatives of LGED and BWDB on operation and maintenance issues. The double membership of these groups provides an instrument of coordination.

Table 4.2. Direct and indirect benefits from FLI Membership.

Field Level Institution (FLI)	Direct and indirect benefits of participation in FLIs
NGO Group	NGO groups are created to improve the social and economic situation of the chars. NGO group members receive health & family planning counselling, micro-credit, and training on disaster management, legal and human rights issues, agriculture, fisheries and poultry/livestock rearing.
Water Management Organization (WMO)	The core task of WMOs is operation and maintenance of water management infrastructure. However, in reality the contributions of WMGs go far beyond that. WMGs have helped in selecting locations for water management infrastructure, and have dug small drains to channel water into khals. Much of the work of the WMGs has involved working with their communities – solving conflicts over waterlogging and issues that arise during project implementation. WMGs also contribute in the resolution of other social conflicts such as discouraging child marriage. WMGs also make an important contribution to disaster preparedness and warning of cyclones. A number of WMGs work alongside school management committees to maintain

	cyclone shelters. WMGs also have a role in liaison with implementing agencies, help in getting access to support from the government and NGOs, and support the implementation of agriculture and forest activities.
Farmers Organization (FO)	FO members have gained knowledge and skills on agricultural technologies through attending FF meetings. For example, they now know how to practise integrated pest management (IPM) and use pheromone traps. As FF members, they have easy access to DAE field officers as well as project directors.
Social Forestry Group (SFG)	SFG members have got the right to 55% of forestry products once the trees are mature. They are also using dry leaves, dead branches and twigs as fuel. Social forestry plantations are protecting the community from different natural disasters and changing climate.
Tube Well User Group (TUG)	CDSP-IV has established DTWs as a safe drinking water source. Each DTW has a care taker family (CTF) which is responsible for the repair and maintenance of the DTW. Thus, the community is establishing ownership of the DTWs.
Labour Contracting Society (LCS)	LCS members (both women and men) are benefiting from short term employment. LCS members are involved in road construction and maintenance, market development, single pit latrine preparation, re-excavation of canals, construction of water management shed etc. LCS members are investing their incomes in gainful on-farm income generating activities (IGAs) like goat and poultry rearing, and also in off-farm IGAs like door-to-door selling of ladies' garments.
Micro-credit Group	This group is the centre of the micro-finance programme as the entire loan proposal, realization, income generating activities of the micro-finance programme and related social awareness activities are implemented through these groups. While selecting group members, special emphasis is given to female-headed households and widows. Group members are trained in farm and non-farm IGAs. Once the members are trained, they invest their micro-credit loans in various IGAs.

One of the examples of how CDSP is building the capacity of the local population to adapt to changing conditions is demonstrated by the fact that Water Management Groups from earlier phases are still functioning. They meet every month to discuss problems and possible solutions, and are taking care of the operation and maintenance of (part of) the infrastructure built by CDSP. A striking example is a WMG in Char Majid (CDSP II), which by themselves constructed and funded a new cross-dam to prevent salinity intrusion. Next to that, discussions with representatives of WMGs revealed that organizations, especially the older ones, have broadened their activities to provide other services to their communities. This included a vaccination programme for cattle, value chain development, and transportation services to markets.

However, several of the WMGs formed by earlier phases of CDSP are not active or traceable anymore. There are a range of explanations for this – significant parts of the CDSP II area have been lost to erosion, with some WMG areas disappearing altogether, and others losing protective embankments along with the sluice gates that they used to operate.

Chapter 5

Role of Women in Development

Nujulee Begum

It has been a conscious strategy in the project to ensure that the consequences of interventions would, as much as possible, benefit men and women equally, and that both men and women would be participants in project planning and implementation. CDSP has made some significant contributions to the empowerment and emancipation of women. This is of crucial importance as women living in the chars are in an extremely dependent and disadvantaged position. They are vulnerable in many ways - not having enough to eat, living in very poor houses (often subject to tidal inundation), at risk from cyclones and storms, and under the constant threat of physical assault from land grabbers and other thugs. Before CDSP, women were rarely involved in planning, implementation and monitoring of development activities.

There are several difficulties in involving women in project activities. These relate to the men of the area, and are encountered at household as well as community level. Women's participation in activities beyond the home compound is seen as a threat to family values in the coastal char areas. The segregation of male and female is one of the fundamental norms governing social organisation of the char communities. Social respectability depends entirely on the women fulfilling traditionally prescribed social roles, and *Pardah* or the veil system has played a part in this. The movement of women outside the homestead requires the wearing of a veil and permission from the household head. Thus, the char women's ability and opportunity to explore the outside world other than their neighbourhood remains limited. Men in the community often express a lack of trust in women's physical and mental capabilities to carry out tasks, and to identify problems and earn cash. In addition, there is an unexpressed fear of losing authority. The non-participation of women in the public sphere leads to a hierarchical ordering of a superior-subordinate relationship between men and women. This relationship is again reflected in the leadership practice that women in society are dependent on men for any type of decision. This notion of dependency on men stems from the woman's ascribed role in the family and society that creates values which lead to a secondary, dependent and disadvantageous position for women in the chars (CDSP II TR17, 2004).

However, due to severe poverty, women had to enter the male sphere of work carrying out such activities as the collection of shrimp fry, plantation of seedlings, construction of thatched houses etc. This has resulted in a change in the traditional gender division of labour in the char areas. To some extent, the women have also exercised decision-making power in the family. This is because during the sowing and harvesting season on the mainland, the majority of the char males travel to the mainland for work leaving the char areas with virtually no male population. This happens more than once in a year, and during these times, the chars remain under the supervision of women. Moreover, due to floods or river erosion during this time, it is left to the women to make the decisions, and take the responsibility to move to another place if necessary. As the male members have to go out to earn enough for their livelihood, the women have to face hard days alone. During this time of male out-migration the entire household activities from earning to expenditure is being governed by the female members of the household. However, other than in these limited aspects, the char areas are an example of strong patriarchal society (CDSP II TR17, 2004).

Another practical difficulty related to women's involvement is their unequal access to training, information, and knowledge, which often results in a limited vision of what they can do or may achieve as women. At the beginning of the project, women were not interested in the planning aspects of the project activities, and usually preferred to be involved in activities that served their practical needs. This meant that they lacked a long-term view of the project and its potential benefits (CDSP II TR17, 2004).

To challenge this attitude and to recognize the importance of involving both men and women in its effort to bring an improvement to the lives of the char population, CDSP has adopted gender concepts that concentrate on ensuring participation of both sexes in all of its activities. CDSP-IV, like previous phases, has aimed to improve the position of women and girls living in the chars. This was both in terms of their practical needs for food, water, shelter, income and support services (especially health), and to the status, security and position of women in their households and in wider society (CDSP II TR17, 2004).

CDSP has adopted a strategy where gender issues are not set apart from its activities but are mainstreamed into them. It was realised that an important aspect of mainstreaming gender issues is by ensuring the active involvement of women and men in the planning and decision-making processes. Thus, the gender strategy is aimed at providing both males and females with equal opportunities and access to knowledge and participation in project activities. This is manifested in the project policy to set as a target a 50-50 male and female membership in Water Management Groups, and a 30% target of women in Farmer Forums. An essential achievement of CDSP has been the provision of legal and secure land titles to landless settlers of chars. An important step forward for women in the land settlement process was the fact that they not only are the legal owner of 50%

of the land allotted to the household, but that their signature comes first on the title document (*khatian*) (CDSP II TR17, 2004).

The challenge then became how to work out practical methods for ensuring women's participation. In this respect, women and men's involvement must go beyond participation in equal numbers as beneficiaries, to achieve active consultation and participation. This enables women as well as men to influence the entire development effort with their respective priorities and expectations. To address this issue, emphasis is given to regular and adequate dissemination of information to women as well as men through their involvement in all field level institutions. It seeks to ensure that any information regarding project activity that is communicated to the men is also made accessible to women. The gender strategy for equal information dissemination is also applicable for information collection. Any information required by the different sectors of the project would be gathered from both male and female settlers in the project area. Female field staff facilitate organisation and communication with the women (CDSP II TR17, 2004).

Some of the impacts of gender empowerment activities of CDSP identified by a recent Gender Impact Assessment in 2018 are described below.

Land ownership

Land titles are granted in the joint names of husband and wife having equal shares, with the wife's name coming first on the title deed. This is exceptional in the context of Bangladesh, where women are generally excluded from ownership of land. This often puts them in an extremely vulnerable, dependent, and disadvantaged position. Not surprisingly, receiving an official title to land is of tremendous importance for women, and has far-reaching positive consequences. The practice of having the woman's name first has now been followed in other places in Noakhali district. Divorce rates, polygamy and violence against women have significantly decreased. This can largely be explained by CDSP's approach of transferring the husband's share of land to the wife in cases of divorce, multiple marriages (without the consent of the first wife) and extreme acts of violence. This can be done as the woman's name comes first on the *khatian* deed. In addition, men are less likely to abandon their wives, as the wives now own half the land and hence are of increased 'economic value' to the husbands. Furthermore, women's social status has increased as a result of their land ownership and they now enjoy somewhat more influence in the home. As women are listed first on the *khatian*, they must be consulted and give their permission when an official loan is being obtained with their land as collateral. Last but not least, women have become increasingly aware of their land rights and are now more vocal. Some even report that they have threatened their husbands with taking legal steps to deprive them of their land in cases of torture and abuse. This has usually resulted in a significant decline or complete disappearance of violence against women.

The efforts made by CDSP-IV to change the lives of women has been recognized by the 2017 IFAD gender award for outstanding results and impact.

Economic condition

CDSP has given women of the char areas better income-earning opportunities. NGOs working for CDSP have provided women with access to saving and micro-credit services, along with training and support for agricultural production (i.e. homestead crops, poultry and livestock), and poultry vaccination, alongside non-farm IGAs such as LCS work, tree caretaking, working as traditional birth attendants, hawking, tailoring, making improved cooking stoves, and cap embroidering. This is alongside the increased income of their husbands. Being the legal owners of land, women are now eligible to take loans from banks using the land titles to secure a mortgage. CDSP studies show that most women report that their overall income has doubled as a result of CDSP activities, which implies a significant improvement in their economic condition and hence their overall position as women. They are now less dependent on men for their survival and of greater economic value to their families. This is of crucial importance within the Bangladeshi context, which is characterized by profound gender inequalities and the dependent, subordinate and disadvantaged position of women.

Fig. 5.1. CDSP IV beneficiary women participating in tailoring training organized by PNGOs.

Food security

The recent Gender Impact Assessment shows that now no household goes without food. Women also reported, that apart from three main meals, they eat homemade and purchased snacks and enjoy different *'pithas'* (rice cakes) during the winter.

Fuel security

Most char households now get firewood from their own trees and purchasing of firewood is now rare. Before CDSP-IV, fuel for cooking was a major problem in the new chars, as all mangroves and other trees had been cleared for habitation. As a result, poor people either had to spend significant amounts of money on wood for fuel, or to cook less frequently (once a day or less). Women spent a lot of time collecting leaves, straw and grass to use as fuel, and it was much more work to tend the fire than when using firewood. This scenario has been changed by social forestry on public land and by household tree plantations producing plentiful supplies of firewood. However, the fuel-efficient stoves distributed by PNGOs do not seem to have caught on. It seems that they do not suit the local food culture and may not be very durable. A few wealthier families are now using gas for cooking – these cylinders are now widely available. Most households in the chars now have solar panels for lighting, fans and charging of mobile phones. This has meant that they no longer use polluting and dangerous kerosene lamps.

Access to healthcare

Although health services are of concern to both men and women, lack of access to health care is particularly difficult for women due to their need for reproductive health services, their responsibility for the care of children, and their relative lack of mobility to travel to services located outside of the CDSP area. Before the start of CDSP, there were no health or family planning services in project char areas. The reproductive health situation of women was alarming, with very high birth rates, and no maternal or neo-natal health care. CDSP has made efforts to improve the health situation in the chars. CDSP-IV supported 13 static and 13 mobile clinics set up by a PNGO and provided training and support via Health and Family Planning Facilitators and Traditional Birth Attendants. For each branch three local women were employed and trained to be Health and Family Planning Facilitators. They made home visits and held group Health Forums advising on family health and nutrition especially for infants. However, these have been limited in scope and may not be sustained after the end of the project. Thus, more work is required in health and family planning services to meet the government's standards for rural health provision in all CDSP areas.

Water and sanitation

Before CDSP interventions, there was a lack of safe drinking water and latrines, which often resulted in epidemic outbreaks of diarrhoea and other diseases. As a result, morbidity and mortality rates were high – especially among the female population and children. Domestic and potable water has been provided from Deep Tube Wells (DTWs) and ponds with sand filters, and household latrines and public toilets have been installed. NGOs have raised the awareness of the population concerning health and hygiene issues. These efforts have contributed to a significant decrease in common diseases like diarrhoea. DTWs have also decreased the time spent by women in fetching water, leaving them with greater opportunities to use their time for income-earning activities.

Education of girls

Before CDSP-IV there were no government or non-government schools in any of the five CDSP-IV chars. The project has constructed cyclone shelters in all the project chars and these are being used as schools during non-emergency periods. However, very few of these schools are being supported by the government and thus lack the resources to provide good quality education. Also, there is need for secondary schools in chars. Nevertheless, it can be generally said that the schools are creating aspirations among children to be educated. This is increasing the number of girls going to primary schools and madrasas.

Chapter 6

Developing the Infrastructure

Mihir Kumar Chakraborty

A core intervention of CDSP has been water management infrastructure to reduce the damage to different crops from salinity, flooding and water logging and improve the environment for crop growth. Effective management of water resources is necessary to enable char dwellers to adapt to climate change, for protection against tidal and storm surges, and to improve drainage. This is done through water management interventions (construction of embankments, drainage sluices, and drainage channels) and by construction of climate resilient internal infrastructure such as cyclone shelters, roads, bridges/culverts, and deep tube wells for water supply. This improves communication, giving people access to outside markets and protection from extreme weather events, and ensures hygienic sanitation and a sustainable solution to the lack of a potable water supply.

Protection from climate change

In char areas sedimentation and riverbank erosion of channels is a considerable problem. There were no water management structures in these chars before CDSP intervention. To address this problem and for protection from climate change, water management infrastructure is constructed by the Bangladesh Water Development Board (BWDB). The following water management activities were undertaken by BWDB:

- **Construction of peripheral embankments (Sea dyke and Interior dyke)**: to stop flooding by saline water during high tides.
- **Construction of dwarf embankments:** to build a hydrological barrier between two polders such as Char Nangulia and Noler Char, as there is difference of elevation.

- **Construction of drainage sluices:** to stop inflow of saline water during high tides and for controlled drainage of the catchment area.
- **Construction of closures:** to close the channels (*khals*) after construction of sluices.
- **Initial excavation of drainage canals:** to improve drainage congestion of agricultural land.

Fig. 6.1. Drainage sluice, Char Nangulia.

Fig. 6.2. Construction of Caring khal closure, Char Nangulia.

During the feasibility study of CDSP-IV, the long-term effect of climate change was considered. However, in the case of some structures such as sluice gates and embankments, a 20 year lifetime was targeted as otherwise the structures would not be cost-effective. This is because the coastal chars are very dynamic and constantly moving. In CDSP-IV, provision is also made for maintenance of water management infrastructure. These works are usually contracted out to contractors and Labour Contracting Societies (LCSs) for smaller items with a quality control function for agency staff. In CDSP-IV, 91 LCS groups with a total of 2,137 members (28% women's participation) were formed to carry out minor construction work including earthworks, road pavements and construction of rural markets. The assessment of LCS (CDSP-IV TR17, 2018) showed that, up to September 2017, there had been 56 LCS contracts with the Local Government Engineering Department (LGED), valued at a total of Tk. 52.4 million. These involved a total of 1,247 men and 381 women, generating a total of 86,991 days of employment. More work has been undertaken since then. The Technical Assistance (TA) team monitors the quality control exercised by the agencies and makes recommendations to the donors on the payment of reimbursement claims. The lion's share of the costs of civil works is covered by Financial Assistance from IFAD, the Netherlands Government, and the Government of Bangladesh.

Four water management development options have been considered in CDSP-IV, from a zero option to a fully-fledged protected area. Due to the highly dynamic character of the morphological processes (continuous accretion and erosion), the preferred option for Caring Char and Urir Char was to avoid any embankments and to concentrate where possible on re-excavation of existing channels and improvement of links between them. This is expected to reduce the area under tidal flooding by about 50%. The impact on drainage congestion and water logging would be even greater.

Fig. 6.3. Lift gate operating deck at C/S of Sluice DS-1, Char Nangulia.

Climate-resilient infrastructure, water supply and sanitation

Climate-resilient internal infrastructure is constructed by Local Government Engineering Department (LGED), including (rural) roads, bridges and culverts, cyclone shelters and *killas* (refuges), *ghats* (jetties) and markets. Climate-resilient internal infrastructure also includes a water supply and sanitation programme by the Department of Public Health Engineering (DPHE) to provide test tube wells, deep tube-wells and single pit latrines. Where possible, works were undertaken by LCSs which channel income directly to some of the poorest men/women of the society.

Fig. 6.4. Female LCS constructing Thanarhat Market, Noler Char.

Rural roads

Rural roads[3] connect villages, farms, markets etc. with feeder roads and embankments. These roads improve communication both inside the area and with the surrounding areas. Construction of rural roads has enormous impact on the economy by giving people access to outside markets, both to sell their own products and to buy necessary goods. To minimize roads crossing existing channels the new road alignments are planned parallel to the existing channels. Thus, the number of bridges and culverts are kept to a minimum. CDSP-IV has constructed 4 bridges (10 m-36 m span) and 81 box culverts to facilitate the smooth running of vehicles and easy flow of water to avoid water logging. For cross drainage provision, pipe culverts have been constructed where necessary. During CDSP-IV, a total of 152 km of surfaced rural roads (BC and HBB) and 134 km of earthen roads have been constructed in the project areas among which 25 km of both types of roads were lost to river erosion.

People of the five new chars suffered from very poor and muddy road communications until the inception of CDSP-IV. CDSP-IV has built roads and

[3] The rural roads are R2 type of LGED Standard, the specifications are: crest width: 3.7 m; side slope: 2 : 1; crest level: 4.5 m (Public Works Department).

bridges that made road transportation comfortable and less time consuming. All kinds of transport can use these roads thus encouraging business and associated sectors to flourish. Char dwellers also reported that due to construction of the bridges, there is no longer a water logging problem and they do not need to use boats and trawlers to cross canals.

Markets

In CDSP-IV, seven markets have been developed, all connected by good roads. However, one market building was lost to erosion in Caring Char. The design is based on LGED's manual for Growth Centre Planning. The markets consist of land raising, construction of multiple sheds (i.e. for vegetable and fish/meat), an open sales platform, internal roads and drains, and a toilet block. A tube well and a women's toilet are also established, with each market having a special raised parking area for trucks.

Better communication and infrastructure have increased access to markets. This ensures that farmers and fishermen can easily sell their products and spend less on transportation. This has also improved access to agricultural inputs, such as HYV and fertilizer, and reduced travel time.

Cyclone shelters

Living on unprotected mudflats only just above mean sea level, char dwellers are extremely vulnerable to tidal flooding, cyclones and other storms. CDSP-IV is working to reduce this vulnerability by building cyclone shelters to give a high degree of security and safety for the people and for livestock by constructing *killas* (raised earth platforms) against tidal bores and cyclones. At the top of each cyclone shelter, there is a stand for hoisting 'disaster' signal flags which can be seen from far distant places to provide the char people with an early warning signal. During the non-emergency period, the shelters are used as primary schools, places of social gathering and as Madrassas (religious education institutions). The shelters are also used by the Union Disaster Management Committees (UDMC) to hold training on disaster preparedness.

CDSP-IV has constructed in total 39 cyclone shelters, among which one shelter in Caring Char was lost as the land where it was built has been eroded and two more in Noler Char and Caring Char are likely to be lost in the same way. Riverbank erosion is a huge constraint in CDSP areas. The total number to be constructed has been determined by planning one cyclone shelter for every 500 households. Shelter sites are selected in densely populated areas preferably near important market places and *Samaj* (community). Approach roads are being built to each shelter to provide easy access during bad weather.

Fig. 6.5. Thanarhat Bazar Multipurpose Cyclone Shelter cum Primary School, Noler Char.

The 2016 Annual Outcome Surveys (AOS) recorded that only 2% of the sample CDSP-IV households reported being displaced due to flood, cyclone, or tornado, and only 3% reported loss of crops due to flood and drought in the previous 12 months. This is considerably less than the 42% reporting displacement and 47% crop loss in the 2011 baseline survey period. BRAC (one of the CDSP-IV partner NGOs) mid-term assessment, carried out in 2016, recorded that 87% of sample CDSP-IV households reported that they had a cyclone shelter near their house, compared to only 9% in the 2012 baseline survey.

Other infrastructure

CDSP-IV has constructed two Union Parishad offices in Char Nangulia for Chandi Union and another in Boyer Char for Hosui Union. Also in Chandi Union office accommodation was built for female staff of government agencies and NGOs. The project also constructed 12 *killas* (refuges for animals) of which 4 were lost to erosion.

Some of the key areas impacted by improved water management are described in the section below.

Reduced flooding and improved drainage

The PRA study found that, with protection from embankments and sluice gates, major flooding is no longer a concern in the CDSP areas. However local people report that drainage is gradually becoming more of a problem as: (i) siltation of drainage canals is reducing their water carrying capacity; (ii) there is siltation downstream of some sluice gates; (iii) there is continued development with construction of minor roads without enough cross-drainage culverts; building of new houses, expansion of markets, construction of ponds with raised banks, and enclosure of land with raised boundaries all create numerous small pockets which get flooded and water-logged during the monsoon season. Better maintenance could reduce the problem of siltation, and Water Management Organizations (WMO) are encouraged to take the initiative to organize such work. WMOs continue to function but their role is often (but not always) limited to sluice gate operation and stopping (or removing) cross dams in canals built by fishers. The responsibility for maintenance is split between WMOs, Union Parishads, BWDB and LGED. Efforts are made to encourage optimal communication and cooperation between these institutions.

Salinity in Land and in Drinking Water

The level of salinity has been significantly improved in CDSP areas, with people reporting to the PRA study that there has been a reduction of 90–95% compared to the pre-project situation. This reduction has been due to embankments built by BWDB, and in many parts of the CDSP areas salinity is no longer seen as a problem at all. However, there are differences between chars depending on their location, with more inland chars showing a greater tendency towards decreased soil salinity. In some locations, drinking water from tube wells is still saline, and lack of fresh groundwater is considered to be the main bottleneck in cultivating high-yielding boro paddy. Extraction of water from the deeper (300 m) aquifer for irrigation is being discouraged, since it is used extensively for drinking water, for which it is the only available suitable and safe source.

Soil Fertility

The PRA survey showed that the soil fertility was as good as other more upland parts of Noakhali district, but the production may be less, due to lack of skills and investment funds, and lack of irrigation facilities in the boro (dry) season. However, with the improvement in drainage and flood protection stemming from work carried out in CDSP IV (which has also reduced soil salinity), farmers are able to take up new crops and farm enterprises and increase production of existing crops. Cropping intensity has increased and farmers are using HYV and hybrid varieties. In the majority of the chars, farmers cultivate two crops and in some areas even three crops a year.

Water supply and sanitation

In order to ensure sustainable livelihoods and quality of life, access to safe drinking water and sanitation is of crucial importance to CDSP. Water-related diseases are the most common cause of illness and death among the poor of developing countries, and this is true in Bangladesh. Diarrhoeal and other water borne intestinal diseases were widespread in char areas. This was due to the lack of safe drinking water in the project areas. People were completely dependent on surface water sources such as nearby ponds, rivers, and ditches. In addition, most of the ponds and ditches fall dry in the dry season. Every year, an epidemic of diarrhoea used to break out and several people used to die from the disease. Aside from suffering, such diseases lead to loss of working days and income, increased expenses, and sometimes result in loss of family assets used for health service expenses. There was a handful of deep tube wells, provided by some NGOs and private initiatives, but the number was insufficient for the vast population living in the char area. People used to travel a great distance to a deep tube well to collect water. Sometimes there were lines of men and women with pitchers at each tube well. Women and girls used to spend a lot of their time collecting water. The baseline survey of CDSP-IV shows that safe drinking water coverage in the area was 46% and sanitary latrine coverage was 6.5%. In addition, there were no health services in the project area.

Nargis Begum, aged 32 years, from Noler Char is a caretaker of a deep tube well provided by CDSP IV. She says, "I feel proud as a caretaker family of a deep tube well sunk near my homestead and I had an opportunity to ensure access to pure drinking water for my neighbours for the last 6-7 years. I can recollect the past days when everybody had to drink dirty and saline canal water or collect water from a well which was normally 2-5 km away and took 1-2 hours walk, with the water collected mainly by women. Now-a-days every household has a sanitary latrine from CDSP IV which has had a huge positive impact on women, as before they had to wait until sunset to respond to nature's call. That was an intolerable situation. Because of all these interventions, water borne diseases are now reduced greatly. These days we seldom use Oral Rehydration Solution (ORS). These significant changes happened only because of CDSP".

Water Supply Service

In CDSP areas, tube wells are now used by almost all households. Currently 98 % of the households in CDSP-IV project areas have access to safe water available within, on average, a 55 m distance. The water is from deep tube wells installed by the project. At the same time, the distance between a household and its water source has also become shorter with the increased availability of tube wells.

Fig. 6.6. Woman collecting drinking water.

One deep tube well is provided for 15 to 20 households within 55 m of household premises and one has been installed for each cyclone shelter. So far, 1,475 DTWs have been installed and among them currently 1,372 tube wells are functioning. 103 DTWs have been lost to erosion. DPHE has worked closely with PNGOs, who have been responsible for forming Tubewell User Groups (TUGs). PNGOs also selected and trained two women from each group as caretakers – who were given tool kits for simple maintenance tasks. In total, 1,532 TUGs were formed with 27,654 female members. They received latrines, and raised awareness of the importance of good hygiene practices. DPHE is rightly proud of achieving "100% total sanitation" in CDSP IV. Due to sinking and installation of DTWs in the project chars, the majority of the population now has access to pure drinking water. In addition, six test tube wells are sunk to get advance information on the aquifers of the project area.

Fig. 6.7. Women are collecting drinking water at the inception stage of the project, Noler Char.

Sanitation Service

CDSP followed the 'one household one latrine' ideology to ensure the minimum sanitation service level. Thus, every household received a hygienic latrine with an appropriate faecal sludge management system. Households are responsible for proper maintenance of the latrines for continual use and improved hygienic practice. Through coordination with DPHE, PNGOs ensured the installation of 25,639 ring-slab latrine sets in project areas, of which 710 latrines were lost to river erosion. Regarding the construction of latrines, as of September 2017, 11 LCS contracts worth Tk. 4.38 million had generated 1,091 days of work for men and 642 for women.

Fig. 6.8. Single pit latrine, where the super-structure is built by the stakeholder and the sub-structure by DPHE, Noler Char.

It has been observed that diarrhoeal diseases have reduced to a minimal level. This result is based on the NGO statistics of saline packet demand and distribution which saw no demand of saline packets (Jan-June 2017), while it was over 200,000 on average during the early days of the project.

Outputs of WATSAN activities

Through its water and sanitation activities, CDSP-IV managed to ensure basic water and sanitation (WATSAN) services to the beneficiaries of the project areas. Access to preventive health services significantly reduces the vulnerability of poor households to illness-induced income erosion and expenditure crises. Some of the key outputs of this sector can be seen in Table 6.1.

Table 6.1. Outputs of WATSAN activities.

Major intervention	**Output/Indicators**
Availability of safe drinking water	98% of households (HHs) have access to safe water available within an average 55 m distance from DTWs installed by the project.
Access to basic sanitation service	74% of HHs have access to an improved sanitary facility.

Major intervention	**Output/Indicators**
Use of soap or ashes in the latrine	79% of HHs have soap and water at a hand washing point inside the latrine.
Hand washing with soap at critical moments	75% of HHs have soap and water at a hand washing point commonly used by family members.
Changes due to safe water and sanitation	Significant reduction of water borne diseases like diarrhoea, hepatitis, scabies etc.

Chapter 7

The Land Settlement Process

Md. Rezaul Karim

The newly accreted land becomes the property of the government (*khas* land) and is transferred to the Forest Department to plant trees that help to stabilize the land. After 20 years, the land is considered as fit for settlement. However, while land is accreting, it is also eroding in other places and it is estimated that each year 26,000 people lose their land through erosion. Without anywhere else to go, many of them try to rebuild their lives on the newly emerged chars, often before the 20 years have expired. They occupy the land illegally and, in some cases, have to 'buy' it from local power brokers. There was previously hardly any administrative action of the government in the new chars. The influential people of the chars used to dominate the area. These so-called *Bahini* (Armed gangs) first distributed the land to the landless people taking money from them, and gradually the *Bahini* were divided into many groups guided by different leaders in different areas. Sometimes fighting took place between them to expand their own area to dominate or to take control of another area. The *Bahini* subsequently expanded their activities in several dimensions. They were trying to act as part of the administration and used to interfere in all sorts of matters of the local people, mitigating the problems in the char area by taking money from both parties. They took money from the people as inducement (*chanda*) regularly and tortured brutally those who were reluctant to pay the inducement. Over time the law and order situation in the project chars became worse. For example, the *Bahini* people used to take cattle from the char dwellers; they abducted teenage girls and raped them and kidnapped the people and demanded money from them. Char people had to obey their commands (CDSP-IV TR6, 2013).

Case profile of Sultan Ahmed (Story of land grabbing)

Nobogram appeared before 1970 and people started living there after its formation. This was a char dominated by the *Bahini*. Sultan Ahmed is from Nobagram. "We were the trespassers settled by the rich absentee land grabbers who claimed land of this char as their ancestral land. There were other people like them in the field and they claimed themselves as *boya* and occupied land claiming ancestral land ownership. There were four contending parties of the *boyas*. We cultivated their land as sharecroppers and got a small piece of land each for housing and farming. We were poor, landless and victims of the river erosion. Thus, we accepted them as our mentors. My mentor was from Kabir hat in the main land. He had 9.00 acres (3.5 hectares) of land here and gave me 1.00 acre (0.4 hectare) of land for my housing and farming. I had draft power and cultivated his land as a sharecropper. He was always ruthless towards me. If I could not produce enough rice due to natural calamities or shortage of inputs, he misbehaved with me as if I were his slave."

"In fact, we were their *lathial* (musclemen, literally those who wield the stick) to protect their land that we cultivated from other contending parties. They forced us to pay them money for the land that we held on different pretexts such as annual revenue, permanent settlement. The *Tahsilder* (government revenue officer) used to come to collect annual land revenue. He contacted the local *neta* (leader), who was the *boya*, not us. The revenue collectors took much more money than the actual revenue. We could say nothing against them."

An essential achievement of CDSP has been the provision of legal and secure land titles to people who are occupying land on the project chars. These people qualify for land titles (a 99-year lease) under the government's policy for distribution of char land. The settlement process follows the Government Khas Land Distribution Policy (1997) that stipulates that one landless household can get not more than 0.6 hectare of land for each household. Getting legal title to the land releases them from the clutches of the land grabbers that exploit their vulnerable position.

Rahena, aged 55 years, from Noler Char says, "I lost my husband in the hands of '*Bahinis*'. The *Bahinis* used to put extreme illegal pressure on us to take our land but we did not surrender. They have forcibly taken five of our cows, four boats, cooking pots, food and other household stuff. Thankfully since CDSP IV started working, we received permanent land settlement by government and now nobody will be able to evict us".

The process of char evolvement is shown below.

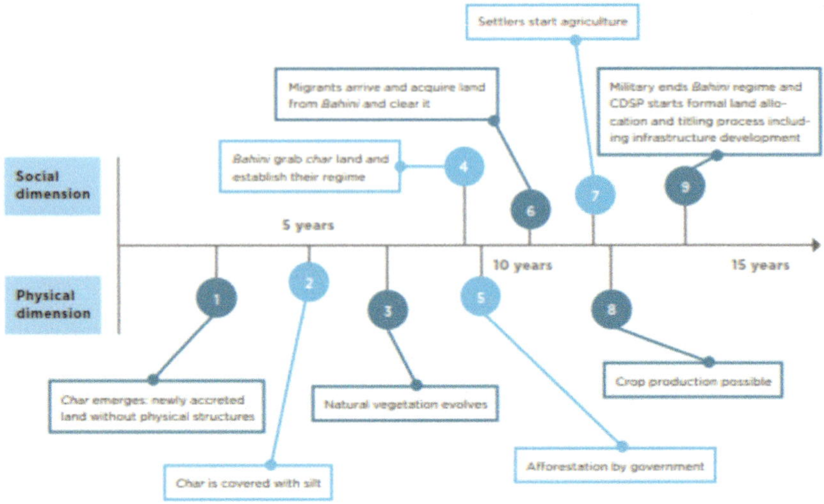

Typical char evolvement. Source: Rosendahl et al (2015)

Fig. 7.1. Mr. Laurent Umans, 1st Secretary, EKN is handing over the land title to the landless Mr. Shahe Alam and his family of Purba Ramahatpur, Noler Char, Hatiya at the location of Saddam bazar, Hatiya, Noakhali, 2015.

Table 7.1. Settlement status of households.

	CDSP-I & II	CDSP-III	CDSP-IV baseline	CDSP-IV
Settlement programme / land title	58	87	1.2	71
Occupying *khas* land	7	8	91	32
Purchased land	42	28	8	6
Inherited land	18	6		1
Sample size (n)	200	200	1400	200

Land titles are granted in the joint names of husband and wife, with the wife's name coming first on the title deed. This has had far-reaching benefits for women which has been described in Chapter 5.

In recognition of its Gender Equality on Land Settlement to the landless families, Char Development and Settlement Project-IV has been awarded IFAD Gender Award for Asia and the Pacific Region for 2017.

Fig. 7.2. Mohammad Chuttu and Nasima Begum of Solaiman bazar, Char Nangulia, Subarnachar Upazila showing their land title after getting the land settlement from CDSP-IV.

With official land title, people of the project areas experience greater economic security and are no longer forced to pay fees to land grabbers. Secure tenure of land encourages people to invest in their property – building better houses, digging fish ponds, cultivating crops, and improving the quality of their land, thus enabling them to improve their economic situation and livelihoods (CDSP III TR7, 2010).

Land settlement under CDSP-IV

The land settlement process under CDSP-IV follows the provisions of agricultural *khas* land management and settlement policy 1997, published by the Ministry of Land, and can be divided into the following broad stages:

1. Plot-to-plot surveys (PTPS)
During a plot-to-plot-survey (PTPS), cartographers measure each plot on the respective char, draw an exact map of the land based on cadastral surveys and note down the details of the inhabitants under the general supervision of the Deputy Commissioner of the District. The maps and the information about the families are then deposited and published in the Upazila (sub-district) Land Office. Complaints against the findings can be submitted within 30 days. Information about the upcoming *khas* land settlement and PTPS is disclosed in the locality through public notice and local meetings which are known as 'Information Dissemination Meetings'.

2. Hearings for Landless Family Selection
The Upazila Committee holds public hearings in the field for disposal of the objections and selection of landless households. While in standard land settlement processes the settlers are requested to travel to the Upazila Land Office to look into the files, CDSP-IV organizes public hearings at the village level. During these hearings, each case is called out to confirm that the family and all listed members are living on the plot. Other participants can object, for example if they know that an applicant owns a plot of land somewhere else. Settlers are assisted in filling up the official forms quickly and correctly. One family can get not more than 0.6 ha (1.5 acres) land according to the government policy.

3. Settlement Case initiations and legal formalities
Once the hearing is concluded, the list with the identified landless households is transferred to the Upazila Land Office which prepares the official resolution of the meeting and the Assistant Commissioner (Land) officially initiates the settlement cases (*Jamabandi*) for each family according to the resolution. District Committee approves the list and the Deputy Commissioner approves the *Jamabondi* cases and sends these cases back to the Assistant Commissioner (AC)

(Land). After receiving approval from the district level, the *Kabuliat* (deed of agreement) has to be signed by both the selected landless households and the land authority. Under CDSP, *Kabuliat* signing by the AC (Land) and the deed registration by the Sub-Registrar are done at the village level due to a special arrangement (in other cases this is done at the Upazila Offices), which saves the families time and costs for the often-difficult travel to the Upazila Office.

4. *Khatian* (Land Title) distribution

Once registered, the details of the settlement cases are entered into the land database of CDSP-IV. The project has developed a Land Records Management System (LRMS), which allows record keeping of every land attribution and helps to prevent double assignments. The LRMS produces computerized *Khatians* (Land Titles). As the last step in the process, *Khatian* distribution ceremonies are arranged to hand over the land titles, making the selected family the owner of the respective plot on a permanent basis. The *Kabuliyat* (Lease Deed) is valid for a period of 99 years. The land, once allocated, cannot be sold and only be transferred by inheritance. The beneficiary family can construct a house on 0.03 ha of land with the remaining land strictly reserved for agricultural purposes.

Process Innovations

Under CDSP, a number of innovations, such as the Plot-to-plot surveys, information dissemination meetings, the public hearings, the *Kabuliat* signing and registration of deeds on village basis, and the Land Records Management System, have been introduced to the land settlement process. The aim of this is to make records less vulnerable to improper alteration and more accessible to the public. Online based modified LRMS software is now running in the district and Upazila level project Land offices. A server computer with required IP Address has been established and installed in the PD Office (Deputy Commissioners' Office) of MoL, Noakhali.

Another change that CDSP introduced to the process is improving the position of women regarding land rights. The wife's name is now written first in the legal document. As a result, she is legally entitled to 50 percent of the owned land. This strengthens her position in the family, provides her uninterrupted access to the land and a legal position in many decision-making processes. For example, if the family wants to use the land as a collateral for credit. Also, if the husband should abuse his wife or it is proven that he is involved in illegal activities, legal steps against him can now result in him losing his share of land.

These elements are unique to land settlement under CDSP. Over the years of CDSP implementation the process has been streamlined and simplified making it faster and more accessible for the char settlers.

Minimum time needed: PTPS to Title distribution

The land settlement process is a time consuming, multi-staged complex work. A minimum of 14 months' time is needed to complete a settlement case and distribution of land title. Many upazila and district level offices and committees are involved in this process. Procedure and CDSP practice to complete a land title (*Khatian*) and required minimum time frame is stated below:

Plot to plot survey and map drawing
⇓ *2 months* ⇓
Occupying list preparation
⇓ *1.5 months* ⇓
Map inking
⇓ *0.5 months* ⇓
Map checked & approved by UNO/DC
⇓ *2 months* ⇓
Occupying list publication
⇓ *1 month* ⇓
Hearing & LLF selection
⇓ *2 months* ⇓
Resolution preparation and approval by UZ & District committees
⇓ *2 months* ⇓
Kabuliat (Deed of Agreement signing)
⇓ *1 month* ⇓
Deed Registration
⇓ *1 month* ⇓
Land title preparation and distribution
⇓ *1 month* ⇓
=14 months.

Table 7.2. Land Settlement Achievement under CDSP concept (CDSP-I to CDSP-IV).

Project Phase	Settled Land (Acres)	Total Beneficiaries (HHs)
CDSP-I (1994-2000)	5,842.00	4,494
CDSP-II (2000-2005)	10,188.00	7,837
CDSP-III (2005-2011)	10,820.00	8,323
Sub Total	26,850.00	20,654
CDSP-IV (2011-2018) (Progress till Dec 2017)	15,527.00	11,944
Total	**69,227.00**	**53,252**

Note: The number of female and male beneficiaries are equal. Husband and wife share equal (50:50) ownership of the settled land.

Since inception of CDSP-IV, the project has successfully handed over *khatians* to more than 12,000 households for over 6,300 ha of land (CDSP-IV TR13, 2018).

Chapter 8

The Power of Agriculture

Md. Bazlul Karim

Agriculture is one of the main sources of livelihood for char dwellers. When CDSP started, agriculture was the principal occupation in the project areas but there has been a general decline in the importance of farming as petty trade and day labour increased. The increase in petty trading across all CDSP areas, but, in particular in CDSP-IV, seems to be due to improved communications and markets. Jobs in services, along with driving (especially CNG run vehicles), is also an increasing trend across all CDSP areas (AOS CDSP-IV, 2017).

Unprotected coastal chars are vulnerable to regular flooding and have high soil salinity levels due to regular flooding by sea water. This prevents intensive agriculture for year-round crop growth. In most of the unprotected lands along the coast only one rice crop is possible in the Aman season (generally cultivated in December-January), when heavy rains temporarily decrease the salinity of the higher layers of the soil.

However, there is a risk of damage through flooding and drainage congestion. In the winter season an additional crop is harvested in some places, but most of the land is too saline for agriculture.

To address these issues and to promote advanced knowledge on different agricultural activities, CDSP in collaboration with the Department of Agricultural Extension (DAE) works on agricultural development in char areas. While DAE aimed at developing field crops, the partner NGOs (PNGOs) focused on homestead agriculture. The Agriculture Development Programme of CDSP has the following objectives to achieve during its implementation period:

- to increase crop production through high-yielding variety (HYV) cultivation and to improve cropping practices in the char areas;
- to improve cropping patterns and increase cropping intensity with diversification of crops in the char areas;
- to select appropriate technologies for coastal chars through adaptive trials;

- to improve livelihood conditions by economic development through improved agricultural activities for the char inhabitants, with special reference to the poorest segment of the population.

These objectives contribute significantly to CDSP's overall goal to have 'Reduced poverty and hunger for poor people living on newly accreted coastal chars'.

Support for field crops, homestead agriculture and agroforestry

The main activities and targets of the programme are focused on training, demonstrations, and adaptive research. The DAE staff provides training (e.g. on different crop production technologies), field crop demonstrations, arranges field days and motivational tours for farmers to the potential areas, and undertakes adaptive trials along with some field test activities on the coastal chars. These activities are implemented in line with the Agricultural Extension Policy of 1996, through the formation of Farmers Forums (FF) which is a core group approach in extension services. In CDSP-IV, with an average number of 60 members per group 90 FFs have been formed through which 5,400 farmers have been reached with 42% female representatives from 29,000 households. The remaining households were reached and motivated through other activities such as demonstration plots and field days.

In CDSP-IV, the DAE formed 90 FFs at mass meetings each with 60 members and 42% female representatives. The total FF members are 5,400 which is about 20% of the total number of farmers in the project area. FF members built their knowledge and good practices through regular monthly meetings with DAE staff. All members participated in one-day training sessions and 1,900 farmers participated in a four-day residential training covering a range of seasonal topics and visited different research stations and seed farms for hands-on training. In addition, 13 Training-of-Trainers courses were organized for DAE and PNGO staff. 72 farmer motivational tours were organized to learn different techniques and new technologies in the region. A total of 1,080 demonstration plots showed farmers new technologies and new varieties and 84 field days were organized around these demonstrations (IFAD, 2018).

Cropping in all CDSP areas is dominated by paddy, which is cultivated by over 90% of farmers. The paddy is predominantly rain fed transplanted Aman, with very little Aus (July-August) now being grown. Boro (March-May) is becoming significant in CDSP areas. This expansion of Boro is driven mainly by the adoption of hybrid seeds. Farmers have been investing considerable sums in irrigation - sinking tube-wells to a considerable depth which can result in over-abstraction, posing a threat to fresh water supplies for domestic use, and making irrigation non-sustainable.

Micro-finance services have been complemented by promoting livelihoods in vegetable gardening, private nurseries, poultry and livestock. One of the major changes reported over time in all the chars, both in the current and previous CDSP areas, is the transformation of agriculture. Settlers have cleared jungle and now cultivate field crops, including the vegetables which they used to grow in their original homes (mostly South Hatia). Field crops such as rice (Rajashail, a local variety of rice cultivated in the Aman season), soybeans, country beans, chillies, groundnuts, watermelons etc. are the main crops. Okra has also emerged as a major cash crop due to huge demand in Dhaka.

Case Profile of Morjahan Begum: Economic benefit from Plant Nursery

Morjahan Begum, aged 45 years, lives in Char Nangulia with her husband and 13 children (6 sons and 7 daughters). She was a housewife before joining the NGO-SSUS's (PNGO of CDSP IV) development work. The economic condition of Morjahan's family has improved a lot since she got involved with the PNGO development process.

Initially, she borrowed Tk. 2000 as microcredit from SSUS to establish a plant nursery. Recently, she borrowed Tk. 35,000 as microcredit for the expansion of her nursery. Seasonally, she earns from Tk. 60,000 to Tk. 70,000 from selling seedlings to neighbours and farmers in the local market. In the current season, they have supplied 800 quality seedlings to SSUS at a cost of Tk. 8,000. In 2014 she bought a motorized carrier called "Nosimon" an auto-van at a cost of Tk. 80,000, using the profit from her nursery operation, for transporting seedlings and other crops to distant locations. During harvesting period, they earn about Tk. 10,000 to Tk. 15,000 per month by renting out the "Nosimon" to others.

In addition, Morjahan helps her husband in farming and processing of crops. She also purchased a paddy thresher at a cost of Tk. 5,500 for paddy threshing. They also rent out the machine to others at the rate of Tk. 200 per day.

Morjahan and her husband also rear poultry and cows. They earn about Tk. 10,000 per year from selling ducks (250 Tk./pair).

The PNGOs focused on items not commonly cultivated in the chars - cauliflower, cabbage, brinjal and long beans, and that can be produced in homesteads or on small pieces of land during the winter season. These crops only need a small quantity of irrigation water, which is normally provided from the pond dug by almost all households to obtain soil to raise their homesteads above flood level. More than 10,000 women group members have been trained in homestead vegetable and fruit crops and 125 women group members in nursery plant production. Being members of NGO groups, they have access to microcredit for farm-IGAs like nursery establishment.

After receiving training on cultivation of homestead vegetables and fruits, the project beneficiaries and farmers are now producing a huge quantity of vegetables for their family consumption as well as for selling in the retail and wholesale markets. Different HYV/hybrid rice and vegetables are demonstrated in the farmers' plots and technologies have been transferred through 1-day field based and 4-day residential training. In the project period more and more rice and vegetables are cultivated; crop yields have gradually increased from 4.0 to 6.0 t/ha. The overall cropping intensity has increased from 127% to 194% by 2016 which is similar to the other areas of the country. One of the reasons for this increased production is due to the decline in soil salinity over the years as the protection of the area through new infrastructure is improved. In addition, project area beneficiaries were trained on Vermi compost and are now using it in their fields. Some farmers are even earning money by selling the Vermi compost and worms in the project area.

However, issues such as lack of salt tolerant varieties and shortage of irrigation water need to be solved to fully harness the potential of agriculture in the present as well as in the future proposed project areas.

Due to the anticipated effects of climate change in the coastal chars, it is a priority to undertake adaptive research aimed at testing a range of field and home garden crop varieties for salinity tolerance.

Fig. 8.1. Farmar Bahar Uddin in his paddy trial plot in CDSP-IV, Char Ziauddin.

Sorjon method of vegetable cultivation

The Sorjon method was introduced during earlier phases of the project for production of vegetable crops during the dry period i.e. in March-May with high salinity, and in deeply flooded areas (60-90 cm) during the monsoon. Under normal circumstances, these fields are not suitable for rice and non-rice crop production except for some local traditional low yielding transplanted Aman rice varieties. Sorjon is an Indonesian indigenous technology and involves making alternate rows of ridges and furrows. Since the introduction of this technology in the CDSP-IV area, farmers are now practising this method intensively for species like cucumber, bitter gourd etc. grown on the ridges, and short duration fish culture is practised in the furrows. The technology is becoming popular in Char Nangulia, Noler Char and Urir Char. In the near future, after implementation of all the interventions of CDSP the situation may be changed in Char Ziauddin, Char Nangulia and Noler Char but, due to lack of protection in Caring Char and Urir Char the technology can play an important role for the farmers of these areas.

Fig. 8.2. Sorjon method of vegetable cultivation in Chon Khola, Char Nangulia.

Case Profile of Md Shiraz

16 years ago, Md Shiraz migrated to char Nangulia from Bhola. He is a member of a CDSP-IV Farmers Forum where he learned about Sorjon from the Department of Agricultural Extension (DAE). Shiraz was the first farmer to develop a Sorjon plot in his village. He started in 2011 by developing one acre (0.8 ha) of land and now has three acres, including one acre of rented land.

Initial land development cost Tk. 50,000 to Tk. 60,000 per acre, but five members of his household did this work over one month. Each year the embankments and ditches need to be repaired. Initially these repairs would incur a labour cost of Tk. 20,000 per acre, but this gradually reduced to around Tk. 10,000 per acre after six years. The poles (mainly bamboo), string and netting to support the crop cost Tk. 30,000 to Tk. 35,000 per acre, with the annual replacement cost coming to about half of the initial cost. In general, cultivating the Sorjon crops, including seeds, fertiliser, FYM and pest control costs Tk. 15,000 to Tk. 20,000 per acre per year.

Like most other Sorjon farmers, Mr Shiraz grows cucumbers, various types of gourd, and country and yard long beans. Income from cucumbers is at least Tk. 100,000 per acre which covers all his costs for the entire year, so the income from other crops - Tk. 150,000 to Tk. 200,000 per acre - is all profit.

The ditches hold 60 to 90 cm of water for five to six months and are stocked with different types of fish like carp, sharputi and tilapia, with these being moved to a pond when the sorjon ditches dry up. A net income of Tk. 20-30,000 per acre comes from this fish cultivation. Md Shiraz has used a range of pest control methods including pheromone traps (to control fruit fly in gourds), and sprays made of soap solution, neem leaf and cow urine. However, he says he needs chemical pesticides to control pests in the rainy season, with between 5 and 10 sprays each costing Tk. 300 per acre. In particular country beans are sprayed against caterpillar and aphids.

Marketing is not a problem, with traders coming to the farm gate to purchase his crop. Previously about 4 to 5 years ago Md Shiraz needed to take the crops to the market himself, but now with improved road communication, the traders come to him. He bargains with a number of traders to get the best price. Normally he aims to get the wholesale market price less transport cost (which is Tk. 2-3 per kg to Noakhali market).

The only problem he faced with Sorjon was due to cyclones and flash floods which washed away the growing crop. This may be avoided if there was better drainage in this part of char Nangulia.

Mr Shiraz has re-invested his profits from Sorjon in his farm, buying a power tiller that cost Tk. 120,000, He says that he has planted around 2,000 fruit and timber trees on his land, and he also rents a shop where he sells farm inputs. This shop is managed by one of his four sons and generates a net profit of Tk. 150,000 per year.

All in all, with the new water management infrastructure, farming has become more productive and less risky, while new roads have improved market access, allowing a move from subsistence to commercial farming. Aman (monsoon) paddy remains the main rice crop, and there has been a gradual switch from local to high yielding varieties. The Rabi (winter) season is now the main time for cash crop production. Some Rabi crops, such as country beans, were grown prior to the start of CDSP, but are now grown by 95% of households. Major crops include soybeans, lady finger and brinjal.

Poultry and Livestock

Poultry and livestock rearing are among the important income generating sources of the char dwellers. This is mainly due to the availability of grazing land in char areas. Char dwellers either own or rear through a sharing system where rich people from towns give cows to char dwellers for rearing in return for a share of the profits. However, people lacked the proper knowledge of rearing systems and so were following the traditional methods. As there was no poultry and livestock vaccination and treatment facility in the chars, the mortality level (especially for poultry) was very high compared to the main land. Typically, between 50-70% of each clutch of baby chicks/ducklings was lost to disease (i.e. Newcastle/Ranikhet disease and duck plague) due to lack of vaccination services. Another challenge for the char dwellers was marketing. The producers did not get the proper price for poultry due to lack of marketing facilities.

Due to lack of services from other government and non-government organizations, beneficiaries were demanding support from the project. CDSP-IV through its PNGOs started Poultry and Livestock activities in the project area in October 2014. CDSP-IV implemented the poultry and livestock programme to increase the income of households by reducing poultry and livestock mortality and by providing training in proper management of poultry and livestock rearing.

Due to implementation of the poultry and livestock programme in CDSP-IV, project area people have become more aware about improved systems of rearing poultry and livestock, and about vaccination and treatment of animal diseases. About 75% of the households now have access to poultry and livestock vaccination and treatment services. Due to the vaccination programme for poultry and livestock, poultry mortality and prevalence of common diseases of poultry and livestock is less visible in the project area.

Fig. 8.3. Ms. Salina, Adarsha Gram, Janata Bazar, Char Nangulia.

Beneficiaries are getting necessary technical support and veterinary care for their poultry and livestock from skilled persons (Poultry Workers and Paravets). As a result, people's income from the sector is increasing which in turn allows them to invest additional income in different areas like production, family expenses and education for their children. In addition, the poultry and livestock programme is creating livestock based employment opportunities and improving family nutrition through increased consumption of milk, meat and eggs. A linkage has been established between the paravets and Department of Livestock Services (DLS) which helps them to become more technically sound. Paravets are also involved with veterinary medicine businesses with small dispensaries in local bazaars which ensure good quality veterinary products in char areas and provide the paravets with a monthly income of Tk. 6,000 to 10,000 from primary treatment and the medicine business. In addition, milk production has received a boost because of the setting up of a milk chilling plant by a private company called Milk Vita at Char Bata bazaar near Boyer char.

Value Chain Development

The DAE conducted 6 workshops to plan a strategy for value chain development. Through these workshops some products are identified for the value chain which are: country beans, country bean seeds, hybrid cucumbers, hybrid bitter gourds, okra, watermelons, soybeans, bananas, chillies, tomatoes, snake gourds, ribbed and sweet gourds. Several potential up markets were identified (i.e. Sonapur, Maijdee, Choumohini, Laksam, Nimsar, Comilla, Jatrabari, Kawran Bazaar, Sandwip, Companigonj, Feni and Riazuddin market). CDSP-IV linked Farmers' Forums (FF) to "market actors" with the objective of ensuring market access for growing volumes of produce. The market actors were themselves members of the FF, and they received training from CDSP-IV. These market actors purchased produce from: (i) farmers who brought produce to local markets, (ii) other traders who themselves purchased from farmers at the farm gate; and (iii) via contract farming arrangements. Although market channels already existed before CDSP-IV, the volumes of marketed products increased with significant reduction in transport costs with better road communication.

Fig. 8.4. Solaiman Bazar in Char Nangulia.

A recent study by the project shows that marketing of crops is generally not seen as a problem any more with 80% of FFs reported facing no marketing problem.

Women's participation in agricultural activities

In most of the CDSP-IV households, it was found that women's participation in agricultural and post-harvest activities is very encouraging and significantly more in comparison to other areas of the country. Women farmers are earning an income from plant nurseries which they established with the use of microcredit from PNGOs. They are investing their profit in secondary income generating activities like paddy threshers, 'Nosimon' local carriers and poultry farms. Women are benefiting from a higher level of income and enjoying better livelihood than before joining CDSP-IV.

Chapter 9

Money Matters – Savings and Loans

Kiran Sankar Sarker

Feasibility studies for CDSP-IV have identified access to credit on suitable terms as a key requirement for improved livelihoods and poverty reduction. Once infrastructure development in the CDSP areas is complete, major national Micro Finance Institutes (MFIs) set up their operations and offer microfinance services. However, for the ongoing CDSP project where infrastructure development works are continuing, it was realised that the project should make specific provision to support microfinance for the following reasons:

- There is an immediate need for microfinance that cannot wait until infrastructure is developed and MFI operations are set up.
- There are considerable advantages in linking microfinance to livelihoods (training, demonstrations) and social support (health, hygiene, rights etc). Microfinance groups meet regularly, thus it is a good forum for providing technical training and social support. New technologies can be promoted through these groups by a combination of training and loans for investment.
- Generation of income from microfinance can make char people less dependent on donor funding and allows them to sustain operations after the end of the project.

Group Formation, Microfinance, and Capacity Building

Microfinance activities of CDSP-IV are delivered by four partner NGOs. The selection criteria for selecting MFIs for this purpose are mentioned below:

- Competent MFIs with experience in operating in coastal char areas and commitment to char development.

- Capacity to mobilize adequate capital for offering microfinance services which effectively meant being partners of PKSF[4] which is a financial institution founded by the Government of Bangladesh to finance rural development and provide training on various financial and non-financial services of NGOs.
- Medium size MFIs with Microcredit Regulatory Authority (MRA) licenses and financially viable microfinance operations.
- Institutional capacity to manage microfinance as well as other social and economic development activities.
- Provision of specialized microfinance products that meet the requirement of char dwellers. These include loans for the ultra-poor on more flexible and attractive terms, loans for marginal and small farmers, and seasonal loans. All these types of loan are supported by PKSF.

Human resource and management set up

Each PNGO deploys a coordinator to supervise the total activities in this sector. The higher management of the PNGOs and TA team provides technical and management support. At branch level a manager, six credit officers (COs), accountant and assistant accountant cum office assistant are deployed. The branch manager is responsible for all activities of the social and livelihoods component. Six credit officers are assigned for group formation, loan disbursement, savings and loan installment collection and other project activities such as issues discussed in the group meetings. The accountant is responsible for the microfinance-related accounts, financial transactions in the branch and preparation of the financial report for the organization and TA Team. The assistant accountant is responsible for maintaining accounts, stock and preserving all the documents of microfinance activities.

[4] Palli Karma Sahayak Foundation (PKSF) is a financial institution founded by the Government of Bangladesh to finance rural development and provide training. It is the largest agency for development in rural Bangladesh. PKSF provides financial assistance and institutional development support to appropriate organizations for implementing sustainable inclusive financial programmes for reduction of poverty through creating productive employment opportunities for the moderate and ultra-poor, small and marginal farmers and micro-entrepreneurs.

Base line survey

The staff assigned to microfinance collect base line data in the project area. Data collection covers all households residing permanently in the area. The information is collected using a set of questionnaires jointly prepared by the TA team and PNGOs. PNGOs organize an orientation to train the staff on information collection methods. In addition, a guideline has been prepared on how to fill in the questionnaire. The senior management of the PNGO, TA team member and the concerned manager will supervise and monitor the data collection and compilation process. The PNGO will compile, analyse data and publish the report. The main reasons for the survey are mentioned below:

- To understand the scope of the employment opportunities.
- To understand the present socio-economic status of the area.
- To understand whether it is possible to form and operate a group in the Samaj/Village/Mohallah.
- To justify whether the aim of the program would be reflected in the outcome.

Formation of micro-credit groups

This group is the center of the micro-finance program. The entire loan proposal, realization, income generating activities of the microfinance program and related social awareness activities are implemented through these groups. Group membership is open only to able-bodied females of 18 years or over. During selection of group members, special emphasis is given to female-headed households and widows. A group is formed based on the following structure:

1. There will be 20 to 30 members in each group.
2. In each group, there will be small groups of five members. In each small group, there will be a group leader who will be selected by the members of the small group. The group leader will be the member of the group managing committee.
3. Every group must have a management committee comprising President, Secretary, Cashier and members comprising other small group leaders.

Insurance policies for beneficiaries

Every PNGO has an insurance policy to safeguard the group members. If any micro-credit group member/spouse dies, the spouse gets a financial benefit of Tk. 2,550 from the project. During CDSP-IV a total of Tk. 933,300 death risk benefits were provided to 366 beneficiaries.

Training to staff and beneficiaries

The PNGOs provide the training as specified in the budget and plan. Moreover, the PNGOs provide orientation, discussion on social issues and problems prevailing in the area. To generate income and improve the livelihood of the beneficiaries, group members are trained on farm and non-farm income generating activities (IGAs) (see Chapter 4 for detail). Training is provided by the TA team. Once the members are trained, they invest their micro-credit loans in various IGAs.

Monitoring & supervision

The microfinance coordinator of the credit program is solely responsible for the performance and outcome of the program. He/she regularly visits the field activities. He/she performs specific monitoring according to a checklist and takes necessary measures if there is any gap in the implementation. Moreover, he/she keeps the senior level management informed in case any measure needs to be taken. The coordinator also maintains liaison with the TA team and other stakeholders.

Reporting

All reporting regarding microfinance is done by the PNGOs, TA team and PKSF. Monthly and quarterly reports based on the data base and narrative are prepared.

Output

In CDSP-IV, 26,373 women (94% of the target and over 90% of households) participated in 984 NGO micro-credit groups. PNGOs have provided credit funds through group member savings and their own resources. To develop homestead livelihoods, 21,902 women or 78% were trained in livestock, poultry, vegetable and fish production. 12 paravets and 114 poultry workers were recruited and trained to provide fee earning services. The value chain development scheme trained 13,520 farmers and carried out 9,476 demonstrations (high value crops, pest control, fruit orchards, vermicompost and rainwater collection ponds), along with tours and field days. With specialized training, 125 tree nurseries were developed, with 110 becoming business enterprises. In the non-farm sector 199 women were trained for 30 days in tailoring and given sewing machines. Among those 125 took up tailoring as an IGA, and some of them trained other women.

Chapter 10

Environmental Improvement with Trees

Md. Robiul Islam

To complement the protection provided by the embankments and other infrastructure, CDSP in cooperation with the Forest Department (FD), established protective plantations of trees on mud flats, foreshores and embankments using a social forestry approach. It has been shown that such a protective "green belt" can significantly reduce the damage from cyclones – both to the embankment itself and to the surrounding area. In general, trees provide effective protection from strong winds, whether planted on roadsides or around homesteads. Next to protection from tidal flooding and storm surges, forestry activities – including roadside plantations and social forest block plantation – provide income earning opportunities. Planting mangroves on mudflats also accelerates accretion of new land.

The forestry programme of CDSP is part of the 'Protection from Climate Change' component of the project. Char lands are initially stabilized by the Forest Department by tree plantations, which facilitate sustainable agriculture eventually. The coastline is protected against natural calamities by establishing a green belt on char lands, shores and banks and by maintaining mangroves. In line with the forest policy, social forestry is promoted by giving priority to poorer communities and poorer members of the community in the allocation of contracts for tree plantation. Women and poor people who do not have a land-based source of income are employed on a priority basis in nurseries, plantations, forest management, harvesting and industrial work.

The Forest Department carries the responsibility for all the plantation activities in the chars, except on homesteads. The support for homestead forestry and private tree nurseries is coordinated by the four partner NGOs which is described in Chapter 8.

Social forestry by Forest Department

Different types of social forestry programmes are included in CDSP IV:

1) **Homestead plantations** for physical protection, income generation, and production of fuel wood to alleviate the severe fuel shortage that exists in the project areas;
2) **Embankment, roadside, canal, foreshore and block plantation** for the protection of inhabitants and internal infrastructure from cyclonic surge and storm as well as for the development of an environmentally friendly eco-system;
3) **Mangrove forest management** protects chars from storms and cyclones and helps in the development of newly accreted chars.

- 7,400 ha of mangrove plantation,
- 268 km roadside plantation,
- 95 institutional plantation places,
- 16 killa planting,
- 200 ha foreshore plantation on dyke,
- Embankment planting, non-mangrove block plantation, and canal side plantation are still in progress with achievements of 70% (35 km), 87% (87 ha), and 53% (80km) respectively.

Fig. 10.1. Roadside tree plantation in Selim Bazar Road, Char Nangulia.

Fig. 10.2. Roadside tree plantation in Tegachia Bazar Road, Boyer Char.

In different ways and degrees, all these different types of plantation contribute to the objectives of physical protection and at the same time strengthen livelihood of the char dwellers.

Workshops and training are organized by the Social Forestry Program for staff of the Forest Department, NGOs, Social Forestry Groups (SFGs), Water Management Groups (WMGs), Local Government Institutions and the public. Additionally, the programme organizes periodically information and awareness campaigns on a variety of topics.

In recognition of the very successful plantations, Noakhali Coastal Forest Division (CDSP-IV FD part) programme has been awarded the best tree plantation of the year award in 2017 by Honourable Prime Minister Sheikh Hasina.

Formation of Social Forestry Groups (SFGs)

In CDSP-IV plantations are established and maintained by the SFGs. These groups are the core of the social forestry approach. Such groups are formed for every 2.0 km of roadside plantation, 1-1.5 km of embankment plantation, 1.0 km of canal plantation, 10 ha[5] of foreshore plantation and 25 ha of mangrove plantation. Each group has about 20-25 members, with roughly 60% men and 40% women. The formation itself is implemented by a team consisting of staff of the Forest Department. Until today a total of 630 SFG has been formed with 25 members each, and more than 40% are women. Due to the serious erosion, 49 groups have lost their trees, in particular the 3,800 ha mangrove plantation in Caring Char. Other losses include 65 ha of foreshore planting, 35 km of block plantations, and 44 km of road and canal-side planting.

Before the formation of SFGs is taken up, the population of the different chars is informed about the social forestry programme through mass meetings and motivational workshops. The beneficiaries are selected, and they form a Social Forestry Group themselves. Generally, the beneficiaries are selected from the local people living within one km of the social forestry area; the following people get priority in the beneficiary selection process:

- the landless;
- land owners with less than 0.20 ha land;
- widowed, separated or destitute women;
- families affected by the Social Forestry Program; and
- disadvantaged people, freedom fighters or their children.

Once a SFG is formed there are certain duties and responsibilities the participants are expected to take on. These include:

- participate in the planning process of social forestry management;
- forestry planning preparation and implementation in collaboration with the Forest Department;
- perform duties on the tree plantations;
- perform pruning and training activities as per the FD approved planning; and
- attend social forestry meetings.

[5] A hectare is equal to 1/100th (0.01) of a square km.

Fig. 10.3. SFG member taking care of roadside tree plantation near Mohammadia Bazar, Char Ziauddin.

The new SFG will receive a percentage of the profit raised from the social forestry activities.

Before or immediately after the plantation, a land lease and benefit sharing agreement between the respective land-owning agencies (like LGED or

BWDB, UP), SFG members and Forest Department are organized. These agreements create a sense of ownership and ensure settlers the usufruct rights on government land and trees. The agreements describe the rights, responsibilities, and short- and long-term benefits of the SFG members. All short-term benefits (i.e. intercropping, pruning) go to the SFGs. The long-term benefits are the revenues at the end of the period of the agreement, by selling the timber of the trees once they are matured. The agreements have a duration of 10 years.

Capacity building

Training courses and workshops are key components of the social forestry activities. The training courses covered a wide range of intertwined social issues and practical skills. The participants were SFG members, other FLI (WMO and FO) members, Forest Department staff, staff from other agencies and NGOs. The emphasis of the training programme provided by the Forest Department is on social aspects of social forestry, its approach and methodology, pest and disease management, and plantation establishment and management. In addition to growing trees, social forestry is also a vehicle to achieve socio-economic development (self-reliance, additional income stream).

Given the importance of afforestation in the process of coping with the increasing consequences of climate change, it is essential that the knowledge and information this Social Forestry Program carries is widely distributed among the coastal population, the Forest Department and the involved NGOs. Large-scale campaigns and training are applied to assist in achieving this goal.

Chapter 11

Income and Quality of Life

Kiran Sankar Sarker

Increased income and changes in sources of income

Annual household income

CDSP projects have built and constructed many climate resilient structures in newly developed coastal chars. These include embankments to control floods and intrusion of saline water, drainage sluices to remove water-logging, markets to facilitate marketing and value chain for the products grown by farmers. In general char dwellers are using 30% of their own land (average 1.3 acres) as homestead for housing and homestead gardening with 70% of the land converting to productive farming e.g. 'sorjon', plots, fish ponds, and agricultural land. Having trained on different income generation interventions, they are now diversifying their income sources and, as a result, multiplying household incomes. The average household income for CDSP phases is shown in Table 11.1. The magnitude of household income increase has been confirmed by tracking annual outcome surveys and PCR stakeholder workshops where the farmers said their incomes have increased by three or four times.

Table 11.1. Average household income for CDSP phases.

Income source	Annual income Taka				CDSP IV increase
	CDSP IV Baseline	CDSP I &II	CDSP III	CDSP IV	
Total farm	25,998	93,515	85,369	118,473	356%
Total non-farm	45,952	214,989	223,349	178,452	288%
Total	71,950	308,504	308,718	296,925	313%

Sources: Baseline 2011, AOS 2017 (CDSP-I&II, CDSP-III) and Impact 2018 (CDSP-IV), PCR CDSP-IV 2018

- Income from farm and non-farm enterprises is estimated as being net of enterprise operating costs.
- Average income in Taka is average for all sample households, not just the households with that income source.
- Main farm income includes agriculture (field crops and homestead vegetables), aquaculture, livestock, poultry and forestry/trees.
- Non-farm includes jobs and daily labour, petty trade, business, skilled work, handicrafts, and pension and social income.

(Source: Annual Outcome Survey 2017, TR-12, pp. 14.)

Increased Land Ownership and Housing

Increased land ownership

Providing land to landless char dwellers is the unique feature of CDSP. A total of 33,522 households has received land titles covering 42,578 acres (17,238 ha) during CDSP phases I, II, III and IV (Table 11.2). Land ownership has been given with an equal share (50%) to husband and wife, ensuring wife's name is in the first place on the title deeds. In case of women who are widows, 100% ownership is given.

Table 11.2. Distribution of land settlements with areas and benefited families.

CDSP Phase	Land Settlement (acre)*	No. of Families Benefited with Land Titles (khatian)
CDSP I (1994-2000)	5,842	4,494
CDSP II (2000-2005)	10,188	7,837
CDSP III (2005-2011)	10,820	8,323
CDSP IV (2011-2018)	15,728	12,868
Total	**42,578**	**33,522**

* *Average per capita settled land= 1.3 acres (0.53 ha) where 1 ha = 2.47 acres*

Average land holding per household is found to be 1.3 acres (0.53 ha). From various different studies like annual outcome survey, impact survey and case studies, it is revealed that char dwellers are developing their land in a productive manner to adopt suitable agricultural practices e.g. making 'sorjon' to produce vegetables and fish, vertical gardens to increase cropped areas, and ponds for fish culture. Besides, they are building durable living houses replacing traditional thatched huts. It is noted that until they receive original land titles they seldom develop durable housing and used to live in traditional thatched huts.

With secure tenure of their land and increased income, many households have invested considerable sums (typically Tk. 100,000) in building better and larger houses. Housing has considerably improved (Table 11.3), with the quality of houses largely catching up with those in the older areas.

Table 11.3. Housing in CDSP IV.

		Baseline 2011	Impact 2017
Average size of house	Square metre	**25.3**	**43.9**
Tin sheet/brick wall	% of households	**13**	**84.1**
Tin sheet roof	% of households	**16**	**82.0**

Source: Technical Report 20: PCR, CDSP IV, pp. 22

Improved water and sanitation

Prior to CDSP IV, the baseline survey of 2011 showed that 99% of char households (27,720) used water of reasonable quality from tube-wells, but they had to go large distances to collect this water (345 m in the dry season, and 418 m in the wet season). Now all 29,000 households have access to 1,531 DTWs installed by CDSP IV at an average distance of 60 m.

The baseline survey showed that in 2011, only 6% of households (1,680) were using hygienic latrines, Now, this has increased to 98% (27,442 households); 91% have received water sealed ring slab latrines from CDSP IV. (Source: Technical Report 20: PCR, CDSP IV, pp. 17)

Increased Asset Value

There has been larger growth in asset value than income. The share of this value by different asset classes is shown in Table 11.4. This shows that the shares of farm and non-farm assets have significantly increased while that for livestock has fallen despite growth in numbers of animals and birds. The main non-farm assets are shops (mainly grocery shops). The main farm assets are trees, timber and ponds. The main household asset is jewellery, although solar systems have also become significant.

Table 11.4. Asset value (Tk.) and share by categories of assets.

Indicators	**Baseline 2011**	**Status as AOS 2017**		**Impact at 2017**
		CDSP-I&II	**CDSP-III**	**CDSP-IV**
Average value per household (Tk.)	**35,162**	**393,873**	**374,242**	**261,485**
Share of value				
• Household assets	21%	19%	18%	19%
• Non-farm enterprises	3%	24%	20%	12%
• Farm assets	13%	34%	47%	41%
• Livestock and poultry	62%	14%	14%	26%
• Other assets	1%	9%	2%	2%
Total	100	100	100	100

Source: AOS 2017, pp. 13 and PCR, CDSP-IV, pp. 21

Improved food security

The improvement in food security is rated highly satisfactory (score 6) by IFAD. The reduction in food shortage and increase in self-sufficiency has been one of the indicators of CDSP project goal. The quality of food consumed has been improved - with more vegetables, eggs, meat and fish being eaten. Over one third of homestead vegetables and fruit are consumed by producers' households, virtually all households keep poultry and consume an average of 200 eggs and 14 birds per year. Impact survey data shows that, on average, each household consumed 44 litres of milk (35% of production) and 80 kg of fish (53% of production). (Source: Technical Report 20, Project Completion Report, CDSP IV, pp. 22)

Increased livelihood capital

An impact assessment on livelihoods of char dwellers has identified positive changes in the five livelihood capitals: natural, human, physical, social and financial. Highlights of these changes and their impacts are:

- Natural capital has been improved - with households getting secure access to land, and with flooding and salinity being reduced (via the development of the physical capital of water control infrastructure). Households have also invested in making land more productive - building fish ponds and fish-vegetable systems, raising land for homesteads and horticulture, and planting trees. As a result, land is now more intensively cropped, and it is possible to grow a greater range of high yielding and high value crops and to cultivate fish ponds.

- Human capital has been strengthened via the development of the capacity, skills and knowledge of char dwellers. This has come about through CDSP IV training programmes, and also via the development of social capital, with community organizations such as Farmers' Forums disseminating information and advice and helping to organize training. This knowledge and new skills have enabled people to take up new livelihoods, such as tailoring, and improve the productivity of crops and livestock.

- Physical capital has been increased via the construction of water management and communications infrastructure. Water management infrastructure has improved the natural capital, and communications infrastructure has improved market access and social connectivity. Cyclone shelters and killas protect people and animals at times of disaster, and the shelters also house schools, while tube-wells and latrines provide households with domestic water and sanitation.

- Social capital has been generated via a range of community institutions and by programmes to empower women and build human rights. These community institutions have in turn supported efforts to get secure access to land and have disseminated knowledge to build human capacity. Water Management Groups operate and maintain the vital water control infrastructure, Tube-well User Groups maintain water supply tube-wells, and NGO groups enable access to financial capital.

- Financial capital - savings and access to micro-credit loans have provided resources for investment in natural and physical capital, and provides a buffer against unexpected health expenses, which along with loans for education, has strengthened human capital.

These five capitals have combined to significantly increase household income and enable households to diversify their livelihoods into new farm and non-farm enterprises. (Source: Technical Report No. 13, Household Impact Assessment using the five capitals of livelihood.)

Chapter 12

Our Pride

It is very encouraging for all members of the project team: Government officers and staff, NGO workers, and the Technical Assistance Team, with their colleagues in Government Ministries, in the International Fund for Agricultural Development and the Embassy of the Kingdom of the Netherlands and, of course, the project beneficiaries and participants to have received special recognition for the project's successes and achievements, as follows.

IFAD Gender Award for Asia and the Pacific Region 2017

In recognition of its Gender Equality on Land Settlement to landless families, the Char Development and Settlement Project-IV has been awarded the IFAD Gender Award for Asia and the Pacific Region for 2017.

Land titles are granted in the joint names of husband and wife, with the wife's name coming first on the title deed. This is exceptional in the context of Bangladesh, where women are generally excluded from ownership of land. This often puts them in an extremely vulnerable, dependent, and disadvantaged position. Receiving an official title to land has far-reaching positive consequences. For example divorce rates, polygamy and violence against women have significantly decreased. Furthermore, women's social status has increased as a result of their land ownership and they now enjoy somewhat more influence in the home. The practice of having the woman's name first has now been followed in other places in Noakhali district.

Fig. 12.1. DTL (middle) of CDSP IV holding the Gender Award, Rome, Italy, 2017.

National Award for Tree Plantation 2017

The Government of Bangladesh has awarded the National Tree Plantation Award 2017, presented by the Honorable Prime Minister Sheikh Hasina, to the Forest Department, Implementing partner of CDSP-IV, in recognition of the very successful plantations. The Forest Department is one of the six government implementing agencies involved and is establishing shelterbelts to protect the coastal chars from storms and cyclones. During 2011 to 2017, the Forest Department has successfully established 7,400 ha of mangrove plantation, 200 ha of foreshore plantations and more than 260 km of roadside plantations. 80 km of plantations have also been established along drainage channels.

Fig. 12.2. Prime Minister Sheikh Hasina handing over the award to Md. Saiful Islam, Forester, Char Bata Beat Office, Coastal Forest Division, Noakhali.

British Expertise Award 2018

CDSP-IV has won the British Expertise award 2018 for international positive social impact. British Expertise International is the leading UK private sector organization for British companies offering professional services internationally. Its membership consists of close to 200 corporate member companies ranging from many of the large UK general consultants to a wide variety of other smaller consulting companies.

Fig. 12.3. British Expertise Award 2018.

Chapter 13

When the Project Leaves

Strengthening the Institutions

In order to ensure sustainability and to build institutional capacity, a paradigm shift in the role of the TA team through an active process of internalization[6] by Government agencies has taken place to prepare for future integrated char development programmes. The internalization process is designed and implemented throughout the project period from the third phase of CDSP, aiming at disseminating the "lessons learned" from previous project phases (CDSP I, II and III) among the six implementing agencies. The process of internalization involves identification of key lessons for each of these agencies and the adoption of these lessons in the processes used by these agencies such as training modules. Lessons are also shared more widely through regular project progress reports, special studies and impact surveys. Information is disseminated via a project website (https://cdsp.org.bd/), newsletters (at project level, IFAD Bangladesh and IFAD Asia Pacific levels) and through workshops and seminars (CDSP III TR 8, 2011). Some of the key areas identified where institutional strengthening was considered crucial are:

- The establishment of a water management organization at the field level and creation of functional relationships between these organizations and the government agencies: the BWDB, the LGED and the Local Government.
- Improvement of the bureaucracy of the Ministry of Lands at District and Upazila levels through staff training and introduction of a computerized system for handling the land-settlement procedures.
- Strengthening of the local NGOs through a separate training and support programme coordinated by BRAC during CDSP III.

---

[6] The working definition of internalization in case of CDSP III was: 'Incorporation of mature experience of CDSP phases (I and II) into the policies and working methods of participating government agencies to strengthen their capacity for future project implementation'.

Furthermore, CDSP II contributed to the development of the ICZM imitational framework through collaboration with the PDO-ICZM and participation in the ICZM forums.

Building and Spreading the Knowledge Base

The experiences with char development in LRP and CDSP have accumulated considerable knowledge on the physical and socio-economic characteristics of the char areas as well as the potentials and constraints in char development. With the establishment of the ICZM framework, the demand for the experiences gained in the coastal areas increased. Since CDSP II, there was room to pay proper attention to increasing the knowledge base in char development as well as to the dissemination of this knowledge. Main activities in this respect have been:

- Systematic monitoring of the effect and impact of the LRP and CDSP interventions;
- Feasibility studies of potentially new intervention areas;
- Studies on specific topics in char development: the fresh water situation and the potential for agriculture;
- Regular data collection on water and soil characteristics; and
- Pilot activities in the areas of land and water engineering.

Dissemination of the acquired knowledge occurred through various channels: (i) results have been published in project reports and (ii) thematic workshops have been conducted on certain topics (land settlement, water management, agriculture); (iii) survey results were shared with the appropriate institutions and (iv) efforts have been made to make the CDSP database compatible with the PDO-ICZM database for easy and uniform access.

However, with the withdrawal of the World Bank from ICZM, the implementation of the ICZM concept lost part of its thrust. It is crucial that the Ministry of Water Resources and the Water Resources Planning Organisation take a lead and follow up the objectives of ICZM.

Linkages with Other Projects

CDSP followed an approach of pursuing open relations with other projects, especially those operating in the same area. The aim of consultation and coordination with other development efforts in the char areas of Noakhali was to avoid duplication of interventions and to mutually support each other's activities, amplifying the results. The reports of the feasibility studies created a planning framework that was followed by other projects as well. CDSP was considered to be the trail blazer for new areas, pulling other activities in its slipstream. This has proven to be very beneficial for the char settlers, making more public goods and services accessible to them in a coordinated manner.

The range of project activities presents an opportunity to participate in a range of networks and activities such as the NGO Drinking Water and Sanitation Forum, the Credit and Development Forum, the Market Development Forum, and various human rights, gender and land rights forums (PCR CDSP III, 2011).

Well-defined exit strategy

CDSP-IV has a well-defined exit strategy that has been shown to work by the char development programme. The various implementing agencies take over responsibility for their different activities and continue to provide the support needed. Although BWDB does not have adequate funding for O&M, evidence from char areas developed over the last 25 years is that polders that were developed continue to provide a favourable environment for agriculture and have developed into a zone of prosperous agriculture with considerable areas of high value cash crops. For major/periodic maintenance BWDB has agreed to take part and be involved. For this purpose, Water Management Groups are registered under BWDB and linked to it. For major/periodic maintenance of roads, bridges, culverts and cyclone shelters LGED will take part and be involved. For this reason, WMGs were also involved in LGED works through Local Contracting Societies (LCSs). Monitoring of CDSP developed areas has continued and shows that salinity levels are still falling. Monitoring also shows that the vast majority of settlers have retained their land so land titling has led to sustainable benefits. Where NGOs have been able to develop viable micro-credit operations, they have also sustained their presence in the area and offer a range of services to their members. In addition, other businesses and public agencies have been attracted by the economic growth. Water Management Organizations also continue to function with only limited support from BWDB.

It is also worth noting that indicators have continued to improve in the CDSP areas, with significant changes since the first round of AOS in 2012. This provides evidence of the sustainability of CDSP interventions.

References

BDP2100 (2018) *Bangladesh Delta Plan 2100 (Volume 1: Strategy)*, General Economics Division, Bangladesh Planning Commission, Ministry of Planning, Government of the People's Republic of Bangladesh.

BRAC (2011) *Impact of Char Development and Settlement Project on Improving the Livelihood of Char Dwellers*, Working Paper No. 17. Available at: http://research.brac.net/workingpapers/red_wp17_new.pdf (Accessed 09 April 2018).

CDSP-II TR 17 (2004) *Summary of Overview of the Gender Aspects In CDSP,* Government of Bangladesh/IFAD/Government of the Netherlands, Technical Report 17.

CDSP-III TR 7 (2010) *Impact of the Char Development and Settlement Projects I, II and III* Government of Bangladesh/IFAD/Government of the Netherlands, Technical Report 7.

CDSP-III PCR (2011), *Project Completion Report* Government of Bangladesh/IFAD/Government of the Netherlands.

CDSP-III TR 8 (2011) *Internalization of CDSP Experience For Capacity Building of Implementing Agencies,* Government of Bangladesh/IFAD/Government of the Netherlands, Technical Report 8.

CDSP-IV (2012) *Design Completion Report – Appraisal,* Government of Bangladesh/IFAD/Government of the Netherlands.

CDSP-IV TR 6 (2013) *Base Line Survey of the Social & Livelihood Support Component* Government of Bangladesh/IFAD/Government of the Netherlands, Technical Report 6.

CDSP-IV TR 13 (2017) *Household Impact Assessment Using the Five Capitals of Livelihood Approach,* Government of Bangladesh/IFAD/Government of the Netherlands, Technical Report 13.

CDSP-IV (2017) *(AOS)* Government of Bangladesh/IFAD/Government of the Netherlands, Annual Outcome Survey.

CDSP-IV PR 14 (2018) Government of Bangladesh/IFAD/Government of the Netherlands, Progress Report 14.

CDSP-IV TR 17 (2018). *Impact of employment in project works on members of Labour Contracting Societies*, Government of Bangladesh/IFAD/Government of the Netherlands, Technical Report 17.

Dewan, C., Buisson, M. C., Mukherji, A. (2014) 'The Imposition of Participation? The Case of Participatory Water Management in Coastal Bangladesh', *Water Alternatives,* vol 7, no. 2. Available at:

https://www.researchgate.net/publication/281554817_The_Imposition_of_Participation_The_Case_of_Participatory_Water_Management_in_Coastal_Bangladesh (Accessed 9 July 2018).

IFAD (2018) *Project Completion Review Wrap-up Note.*

Islam, R. (2015) 'To help climate migrants, Bangladesh takes back land from the sea', *Reuters,* 9 September. Available at: https://uk.reuters.com/article/us-bangladesh-clmate-land-idUKKCN0R90U220150909 (Accessed 12 April 2018).

Jashimuddin, M. (2012) 'Forest conservation in Bangladesh: Legal measures and policy support in relation to landscapes and land use issues', *International Law Conference on Conservation of Forests, Wildlife and Ecology*, Kerala, India, December 2012, ResearchGate. Available at: https://www.researchgate.net/publication/283732074_Forest_conservation_in_Bangladesh_Legal_measures_and_policy_support_in_relation_to_landscapes_and_land_use_issues?enrichId=rgreq-5deffeba1263827e4a21339bdb316f2b-XXX&enrichSource=Y292ZXJQYWdlOzI4MzczMjA3NDtBUzo0MDcxNzk0NDMwOTc2MDBAMTQ3NDA5MDU1MjQ3Nw%3D%3D&el=1_x_3&_esc=publicationCoverPdf (Accessed 9 July 2018).

Rosendahl, J., Rahman, M., Rabbi, S.E., Jenkins, A. (2015) *Environmental and Social Vulnerabilities of the Poor under Climate Change Conditions: The Char Lands in Bangladesh.* In M.A. Zanella, J. Rosendahl and J. Weigelt (eds) *Pro-poor Resource Governance under Changing Climates. Addressing vulnerabilities in rural Bangladesh, Bolivia, Brazil, Burkina Faso, Ecuador and India.* Institute for Advanced Sustainability Studies, Potsdam.

Wilde, K. de. (ed) (2011) *Moving Coastlines: Emergence and Use of Land in the Ganges-Brahmaputra-Meghna Estuary.* The University Press Limited, Dhaka.

Wilde, K. de. (Author, Editor), Latif, Md. A. (Author), Groot, J. (Author), Mutsaers, H. (Author) (2000) *Out of Periphery: Development of coastal chars in Southeastern Bangladesh.* The University Press Limited, Dhaka.